Does Science Confirm The Bible?

By Steve Preston

1st Edition

Contents

Introduction

There is no wonder people have so much confusion today concerning religion. You read the Bible and it said some pretty weird things compared to what you were taught in school and you possibly thought this book just can't be true. The quasi-science that you learned in school claimed death terminated all life, that spirits were something made up in fantasy, and that our universe includes everything. They claimed the ability to disappear, healing someone with your hands, reincarnation, demon possession, turning water to blood, or walking on water were all impossible. They told you undirected evolution turned a snail into a person that your reality never could change. On and on I could go as our schools forgot to look into real science from Albert Einstein, Neils Bohr, Edwin Hubble, and Milo Wolfe or just about any reasonable physicist to teach our kids truth. I don't know if this is being done on purpose to eliminate Judeo-Christian beliefs or the teachers are simply stupid and have never heard about the scientists that have brought us the following:

Law of Entropy	Conservation of Matter	Forensic Science
Hubble's Law	Anthropology	Organic Chemistry
Big Bang Theory	Paleontology	Conservation of Time
Conserv. of Energy	Micro Biology	Natural Genetics
Relativistic Physics	Bio-genetics	Applied Epigenetics
Participatory	Haplotype Tracking	Parapsychology
Anthropic Physics	Aerodynamics	Regression Hypothesis
Vibrational Physics	Astrophysics	Xenoglossy
Bio-photonics	Geophysics	Geologic History
The Hutchison Effect	Electro-Plasmology	Transgenics

We will touch on all of them to see if the Bible is telling us the truth. As a hint we will find that there is no mystery, nor unexplained miracle, no scientific inconsistency, no lies, and no reason to distrust the details provided in the Bible. This book joins what we call science and what we call Judeo-Christian religion together by simply letting you know the science of today not by making up stories or hiding things to force some agenda like many teachers must have today. Some want to only believe the Bible and claim science is absurd while others claim the Bible is trash because of what they learned in school. Some even try to consider religion and quasi-science separately for reasonable understanding of each separately. [Ouch!]

Moses wrote about scientific characteristics we only now are beginning to understand completely and Jesus described to his disciples and showed to the world another side of reality we only now can attempt to know. When I speak of Religion in this book I don't mean the religion of "Buddhist multi-god worship", " evil snake biting religions", "Natural Secularism" where man is his own God, or "Muslim moon god worship" where killing Christians assures one of an eternity of debauchery in a lust filled heaven. Certainly many "religions" only work in a fantasy world, but Judeo-Christian religions and the Biblical teaching have proven over and over again that they are based on <u>strong scientific fact</u>. When I say science I am not talking about mumbo-jumbo sciences like Aryanism, Phrenology, Moleoscopy, Physiognomy, what is being thrown around as climatology, or undirected evolution, I'm talking about the repeatedly tested and witnessed sciences. Things that we must discuss not only are the easy questions but also the more difficult descriptions in the Bible and in the various sciences.

What is reality? Einstein was asked, "If a tree fell in the woods and no one was around to hear it, would it make a sound?" His answer was, "There is no tree!" Today's physics uses the hard truth that our reality is totally based on those witnessing it. The Bible helps us understand Physics.

6

What was the Big Bang? Now that we know the Earth is in the center of the Universe according to Hubble's Law. How could the Big Bang have happened in the center of the universe? The Bible helps us understand Physics again.

Where does light energy go? This depressed Einstein for years as he knew there was a finite amount of photonic energy in the universe and if it is directed outward, it eventually would reach the ends of the universe and the world near where we are would go black. The Bible helps us understand Physics.

How could the first age have a day and night when the sun was not created until the 4th day? While part of this is a slight misinterpretation, science helps us understand how really insightful Moses was so long ago.

How was the planet Rahab destroyed? The Bible tells us God simply destroyed it and caused it to shatter. Earlier scientists indicated Rahab or Venus was destroyed eons ago because of something they called the Greenhouse effect. Neither science nor the Bible can separately answer us completely, but together, the answer becomes clear.

How can Angel be real? Does science even understand or provide for them? We find that people must live in our linked universe or our own universe could not exist.

How did Cro-Magnon humans simply appear one day? Neither science not the Bible can separately answer us adequately, but together, the answer becomes clear. DNA tells us Cro-Magnon humans did not "Come out of Africa" so from where did Cro-Magnon come? Moses explained it thousands of years ago.

What are scientists even talking about when they say there is a symbiotically linked universe that is absolutely required to keep our universe operational/ What we find is the

7

Bible helps describe it. Our Biblical texts talk about at least 5 separate "linked universes and other text expand the number to at least 7. Mathematicians indicate there are at least that many.

Relativistic Physics tells us time stops at the speed of light and the Bible shows us what that means.

Relativistic Physics tells us at the speed of light one can see the beginning and end of time simultaneously. The Bible helps here again.

Physical evidence and the law of Entropy show us survival of the fittest doesn't work and the Bible helps us understand how "animal mistakes" happened and established what were called unclean animals.

The Bible indicated a number of people walked on water, healed sickness with their hands, and even converted water into wine. Participatory physics, Bio-Photonics, Heart-Brain studies, and Vibrational matter descriptions explain exactly why these things are reasonable.

Physical evidence showed that a little girl could lift a car off her dad to save his life; none of her bones were broken; and her muscle tissues were not torn. The Bible indicated that with "faith" a person could change reality and move the position of a mountain. New sciences explain how all of this is reasonable and proper and experiments have proven it.

The Bible and other documents describe powerful demons that invade people to possess them and that one can demand the demons to leave. New sciences show exactly what all this means.

Quantum mechanics tells us that not only can time disappear, but also space can disappear so that if something disappears in one spot in the universe it immediately will appear somewhere else to assure conservation of energy and matter. The Bible gives us examples of this so we can understand science better.

Science tells us life cannot end and the Bible explains how that can be possible.

The Bible indicated Giants lived before the time when the Earth became void and without form. Science and new physical evidence describes how that could be and that it is easily proven.

The Bible indicates that after a physical death you sleep but no time passes. Relativistic Physics provides why that has to be true.

The Bible indicates "carnal life in this reality" is temporal while real life is outside our bodies. New science shows how this must be true.

The Bible indicates people are made up of three identities Self, Soul, and Spirit. New science confirms this possibility and descriptions of these three entities include the "Ba, Ka, and shadow"; "Ego, Superego, and the Id"; and many other similar ways to "not talk about the spirit and soul".

Can someone really make wine out of water? The Bible indicates Jesus was 100% man. Doesn't that mean people can make wine with water? Sure enough scientists have done similar things in laboratories.

The Bible "excoriated" Transgenic evolutionary changes in animals unless creator God was the "Director". Science describes how undirected or misdirected evolution leads to horrible problems.

Organic Chemistry No Evolution

Speaking of evolution, the Bible and many supporting ancient Judeo-Christian texts indicate that the initial creation of animal types was our creator God. While it leaves open the possibility of others manipulating genetics over the years and in fact defines times when this type of forced evolution was rampant. It wasn't the only place we find a miscommunication or major limitation of a failed theory. We look to Organic Chemistry to test the model of undirected Evolution to find it simply could not be done. While this left a troubling idea that we have a God watching over us and even creating us there was not much else that could account for massive combinations of sugars developing into a human while fighting the constraints of Entropy continually pushing organisms down to their simplest construct. Dr. Charles McCombs, one of the leading Organic Chemists focusing on the origination and continuation of life helps us understand why undirected evolution cannot work. He indicated the following:

There is not a single example of life resulting from a chemical reaction. Proteins and DNA are complicated chemical molecules that are present within our body. Cells which make up the living body contain DNA, the blueprint for all life, and proteins regulating biochemical processes, leading scientists to conclude these components are the cause of life. While it is true that all living bodies have proteins and DNA, so do dead bodies. These chemicals are necessary for life to exist, but they do not "create" life by their presence; they only "maintain" the life that is already present.

He further identified 3 reasons why undirected evolution can take place during the chain building process that would create DNA that could be replicated

Chemical stability *- stable-Amino acids are <u>relatively stable in water and do not react</u> to form proteins in water, so nucleotides do not react to form DNA in a wet environment.*

Chemical reactivity *- When challenged, the laws of chemistry are not able to predict the sequence of DNA chains. Amino acids react according to their reactivity, and not in some specified order necessary for life. Since all of the amino acids have similar reaction rates; they cannot react in any organized way.*

Chemical selectivity *- As many DNA strains try to build "naturally" without a controller, none would even be the same length much less have a precise structure to allow secondary combinations. There is nothing in nature to do the selecting.*

This whole discussion does not limit directed evolution during any period and even provides for the possibility of Transgenic evolution could be accomplished by Gene splicing or other experimental methods, but all require an outside observer/manipulator. We now know "You", in all your glory, are a community of 50 trillion living cells. Each cell is a living individual, a sentient being, that has its own life and functions but interacts with other cells in the nature of a community so do go thinking so highly of yourself and understand you are simply made up of trillions of little animals. While we will discuss more about a control we know you have 46 individual DNA Strings in a cell nucleus [Alive and Dead] and the alive ones talk to other cells using modulated light as the DNA vibrates to modify color. "Are you ok?" one communication might describe while another might say, "I'm infected." The first transmits to a second and third and white blood cells are brought in to save the day. Some of the white blood cell give up their life to keep you healthy all while you are brushing your teeth and the Bile helps us understand some of this. Each strand of DNA contains a sequence of genes. The genes are the blueprints for each of the over 100,000 different kinds of proteins that are the building blocks for making a human body plus there were also genes that didn't make proteins but controlled the other genes. For example, scientists looked at a group of people, scored them on the basis of "happiness" and tried to find out whether there was a gene that was associated with happy people that was not active in unhappy people. Think about this concept. If you had "unhappiness", an "I'm lazy" gene, and a "fat" gene, your life could be disrupted from its maximum potential.----*It's not that simple*. Coding does initiate predisposition, but DNA can certainly be overridden and it is not the thing that actually controls you. As we find out more about the Soul part of us, you will see what I mean. Until we get to that quit being a lazy, fat, unhappy person and let's discover how the Bible is being proven with science every day.

Biblical Truth

What we will see in this book is that modern Science describes Biblical characteristics including those things that we previously thought to be miraculous, simple Biblical error, or fantasy. We will also see Biblical characteristic help prove scientific and mathematical theory. If you have a religion that doesn't agree with science either the religion is wrong or the science is wrong. That being said, just because a teacher proclaims stories to be fantasy does not mean they are as many apparently know little about science. The new and widely understood science of "Haplotyping DNA mutation" is helping to solidify the Bible as a trusted anchor in science and in anthropology so we will look into it. Once you have been introduced to some of the modern sciences that many teachers shied away from we will look at the Biblical testimony and history to see if anything in science is violated in Judeo-Christian beliefs. By the way; when you were told height, length, and depth were the dimensions of our universe, they lied to you. When they told a bunch of sugar on a stump could be hit by lightning and a living entity would be made; they lied. When they told you angels, heaven, faith healing, walking on water, the devil, demons, and hell were all elements of fiction for religious people; they lied.

12 Dimensions?

Today, physicists define everything as vibrating single dimension strings or waves and the universe and everything in it are described that way. On the surface it seems well regulated just like Moses' description of our universe group and our beginnings, but later you start seeing the nuances and how Moses and other Biblical writers were able to describe details of our universe and the entities in extensive detail. Here are the new scientific definitions or characterizations of 12 dimensions of our universe spit into 4 distinct dimensional blocks or dynamos. We will be looking at them in more detail and comparing the Biblical teaching as we go along.

Life Dimensions or our universe-*Soul [potential life] and Spirit vibrate to make self and life force- the characterization of reality, counteraction of loss of reality, and causing the sensation of Life*

Force Dimensions of our universe- *Electricity [potential energy] and Magnetism vibrate to make E-M Force and photons- the characterization of In-waves, counteraction of entropy, and causing the sensation of Light.*

Matter Dimensions of our universe-*Aether [potential matter] and Gravity vibrate to make Matter and nuclear force- the characterization of Out-waves, counteraction of nothingness, and causing the sensation of particles.*

Time-Space Dimensions of our universe- *Infimum [Potential Space] and Time vibrate to make Space-time continuum and the universe - the characterization of existence, counteraction universal collapse, and causing the sensation of the universe.*

I know these don't sound like the old length, width, height you were told were the dimensions of the universe, but now we know everything is not as it seems and the Bible is helping us put perspective in what we are finding.

- The universe is expanding away from Earth as the center.
- Reality is modified continuously depending on how fast one moves through space.

14

- There are multiple past realities and multiple future realities associated with "now".
- <u>People can change reality</u>.
- Space and Time are no longer constants; in fact, time can stop and even go backwards.
- There are a number of universes and <u>we need them to exist</u>.
- The things we thought were made up or in error and miraculous in the Bible are now being understood as they help prove science.

In order to find out the secrets of the Bible, one must first understand what we can call vibrational physics. I don't mean learning thousands of mathematical equations that take pages and pages to solve. Someone else has done all that all we need to do is gain knowledge form the work of many scientists.

Vibrational Physics

Vibrational Matter-To understand the Bible, we must first understand a little physics [Sorry I said the word that makes you cringe so use some other word whenever I use write the nasty thing. How about physique'? Physicists, or Physique-ists, got together one day and decided there was no such thing as matter. Instead, there was only the appearance of matter and matter changed its characteristics by how fast something Einstein called Aether [the potential for matter] vibrated. The faster this Aether vibrates the more gravity it produces. The combination of gravity and Aether makes Nuclear Attraction or matter. As this matter vibrates faster, it gets more gravity and appears heavier so Hydrogen is much slower than god. If these three things vibrate fast enough, they become a black hole or pure gravity. This following chart shows how the dimensional elements of matter react with vibrational changes. Notice that if you have any matter and vibrate it at 60 exahertz, it turns into gold. If you could vibrate items with a touch, you could be like Midas. You could even turn water into blood or wine.

Chart of Particle Vibrations

Name or characteristic	Maximum Wavelength [meters]	Highest Frequency [Hertz]
Aether, black matter	$*1 \times 10^{+10}$	$<30 \times 10^{-3}$
Human hearing	1×10^{4}	20×10^{3}
Fermion [part mass]	$*1 \times 10^{+4}$	30×10^{3}
Boson [smallest mass]	$*1 \times 10^{-0}$	30×10^{7}
Baryon [electron]	$*1 \times 10^{-3}$	30×10^{10}
Hydrogen/1	1×10^{-9}	30×10^{16}
Berylium/9	1×10^{-10}	30×10^{17}
Silicon/28	3.5×10^{-11}	8.5×10^{18}
Zirconium/91	1×10^{-11}	30×10^{18}
Gold/197	5×10^{-12}	60×10^{18}
Meitnerium/270	3.7×10^{-12}	27×10^{19}
Straight Gravity [Black Hole]	smaller	higher

Photonic Force-These same physicists got together and confirmed that what we sometimes call "photonic energy" is almost the same thing as matter except appearance of photonic energy changes its characteristics by how fast something called Electricity [potential electromagnetic energy] is vibrated in association with another dimensional string called Magnetism. The faster this Electricity vibrates with Magnetism the more electro-magnetic force is produced. If they go fast enough, they become pure Magnetism. The following chart shows how the dimensional elements of Photonic Energy react with vibrational changes. Notice that if you have any Photonic Energy vibrating at and vibrate it at 750 Peta-hertz, it turns into a wavelength we can interpret as visible light. Slower vibrations turn it into radio waves and faster vibrations turn it into deadly cosmic rays. [It's all the same thing.]

Chart of Electro-Magnetic Vibrations

Name or characteristic	Maximum Wavelength [meters]	Highest Frequency [Hertz]
Electricity [potential energy]	5×10^{10}	$< 30 \times 10^{-3}$
Brain function**	5×10^{7}	6 to 10
VHF [radio]	1×10^{0}	30×10^{7}
UHF [radio]	1×10^{-1}	30×10^{8}
SHF [radio]	1×10^{-2}	30×10^{9}
EHF [radio]	1×10^{-3}	30×10^{10}
Microwaves	2.5×10^{-4}	12×10^{12}
Infrared [light]	1×10^{-6}	30×10^{13}
Visible light	4×10^{-7}	75×10^{13}
X-rays	1×10^{-8}	30×10^{15}
Gamma Rays	1×10^{-9}	30×10^{16}
Magnetism [pure kinetic energy]	lower	higher

** it is highly likely that these frequencies are simply catalyst for much higher frequencies actually used by our brains to store thoughts and images.

Life Vibrations-Initially, life vibrations were not considered dimensional requirements of our universe, but Schrodinger's

17

cat changed everything [sorry for not explaining this right now] and it was found that a cognizant viewer or viewers "create" what we call reality. Please don't worry about this right now as we will look at it by itself, right now I am simply introducing the real dimensional characteristics of our universe in case your teachers forgot to tell you.   Life experience became dimensional characteristics just like matter and electro-magnetic forces as matter and electro-magnetics can't make reality on their own. Like the other elements of our universe cognizant life is created by vibrating 3 mutually perpendicular vibrational components [to make a dynamo] of life called the Self, the Soul, and the Spirit. Everyone is made of all three of these things. We'll look at that a little later. Like the other two, life changes depending on how fast the union of the soul, self, and spirit vibrate. This next chart shows how the dimensional elements of Life react with vibrational changes. Vibration levels slower than 10^{17} are almost completely Carnal. This means   a person who is debased, totally dependent on what we call reality, and focused on three things: survival, sex, and self. As one increases the vibration, he slowly becomes self-actualized and begins to notice and appreciate other people's feelings and even too have something called selfless love [able to give up your life for another]. The more one removes himself from his selfish nature the higher he vibrates and the more he can control his "reality". If you vibrate fast enough you can leave your body, walk on water, and even move a mountain. Notice that experiencing selfless love even for an instant drives your vibrational level to something like 6 peta-hertz, so if you are feeling a little buzz, you know where it's coming from. As an aside, Buddhist monks practicing chakra meditations indicate that as they get higher in what they call chakra level, they begin to feel a buzz or hum of some kind. It should be known that at the higher vibrational levels one can CHANGE his or her environment which is a very important characteristic to

18

understand before reading the Bible. While Physicists have been toying with all of these requirements for years, some fake scientist I call "Consensus scientists" try to keep all this detail away from the masses.

Chart of Ethereal Vibrations

Name or characteristic	Maximum Wavelength [meters]	Highest Frequency [Hertz]
General Molecular interaction	*1 x10^{+10}	<30 x 10^{-3}
Unaware Life	*1 x10^{+4}	30 x 10^{3}
Life Awareness	*1 x10^{-0}	30 x 10^{7}
Survival	*1 x10^{-3}	30 x 10^{10}
Sex	1 x10^{-9}	30 x 10^{16}
Need for Companionship	1 x10^{-10}	30 x 10^{17}
Need to help others [self actualization]	3.5 x10^{-11}	8.5 x 10^{18}
Selfless Love	5 x10^{-12}	60 x 10^{18}
Universal Understanding	3.7 x10^{-12}	27 x 10^{19}
Insight into external Universe	smaller	higher

No matter how you sense it, increasing your vibrational resonance frequency, increases your power over your environment.

The next component that is needed to understand some of the mysteries of the Bible is something Einstein called his Theory of Relativity.

Relativity

Albert Einstein was pretty smart and it wasn't long before he destroyed the myth of height, length, and width as dimensions. At the speed of light, distance and time go crazy and the so called dimensions disappear. [Oops!] This didn't happen until the 1940s so if your teacher still told you that, she has been sleeping a while. Einstein claimed that as you get closer and closer to the speed of light, you become infinitely long in the direction of travel and all time stops. I think the example below shows you what he meant. If two guys are near each other and one spins near the speed of light for ten years and stops, the spinning one will have not aged a second and he would been invisible until he slowed down.

The spinner had ceased to exist in this reality, but he was still in this universe. That is where Dr. Edwin Hubble came in and noticed that as high speed objects moved away from us they changed their mass and their radiated electromagnetic signatures became slower. This is called red shift as massive yellow stars become smaller in our reality when it moves away from us. Let me show you what actually happens.

Dr. Hubble discovered that if objects moved away from us very fast, the colors that emanate from them shift towards a more reddish color and these "RED SHIFTS" are packetized as if blocks of things are all going the same speed from us. To the object moving away the colors didn't change, but when we watch them the colors are all different. The light changed into something else because the number of vibrating cycles had to stay the same as shown in the following graphic had to get slower for us to see it. I think the next example will show you what Dr. Hubble really found. As we already discussed slower vibration means smaller objects

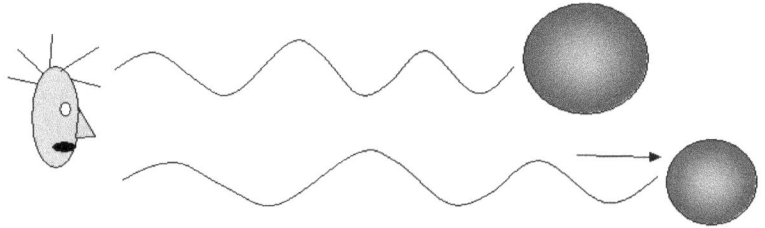

Same number of vibration cycles are created but must be stretched as the object moves away for us to see them.

While this is an accepted characteristic, what it is showing is the universe associated with the object moving away from us and our entire universe stays in sync with 'OUR REALITY" by changing vibrational patterns and vibrational patterns determine matter.

Let me say this again. Time stays in sync by changing matter. The red shift is actually matter being converted to smaller particles in our reality.

What this red-shift means to 'Vibrating" matter is that if we could see the things on the object moving away from us they would **not** be the same things as they would be if there was no relative motion. Let me explain how bizarre this WEL KNOWN phenomenon is.

Lead objects might be gold objects-- or we could "see" radio waves as light-- or any number of things that would make us go crazy. It's best for us not to look.

If the moving away guys could see us, gold objects would be lead and our bodies might be some type of bright blue material or our clear atmosphere may be completely solid. Here is the most important part of this. If a person is moving away from us at near the speed of light, he would gain mass according to Einstein, but he would reduce his mass to lower vibrational elements according to Dr. Hubble. Soon the person you were viewing would be invisible as his vibrations stretched enough to turn him into mostly oxygen [in our reality] rather than mostly water [our at rest body].

Given all that we could say if you go the speed of light, you would turn into Aether in the reality of the viewer.

Let's look at the opposite view. If you are going fast and looking at everything around you and you see a tree. If you are going fast enough, the tree has no width at all in the direction of travel. There is no way the "line" that you see is made of wood and plant cells in your high speed world.

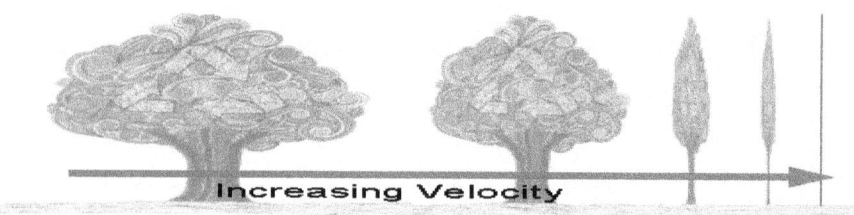

Einstein Confirmed It

If you ever wondered what Einstein's famous formula was all about, let me tell you.

$E=MC^2$ is saying $E/M = C^2$ or <u>Energy and mass are mirrors of each other</u> when there is a differential vibration or speed associated with the speed of light. If our universe begins moving things around from having more energy than mass or vice versa we change our reality.

Dark Matter

That brings us to another anomaly not caused by the Bible. We are told most of the universe is dark matter and dark energy to allow the galaxies to stay galaxies. Dark matter and dark energy are easy to define. They are matter and energy that MUST be, but aren't. This is not religion folks this is science. The "Matter" you can see is now called, baryonic matter [visible matter] and neutrinos [sub atomics that have partial existence]. This dark matter stuff doesn't even interact with electromagnetic radiation, such as light. It is completely invisible to the entire electromagnetic spectrum.----But it's there and there is a bunch of it. Roughly 68% of the universe is dark energy; dark matter makes up about 27%; and the rest - everything on Earth, everything ever observed with all of our instruments, all normal matter - adds up to less than 5% of the universe. Scientists found the stuff by looking through the Hubble telescope at objects way out there. In 1998 a supernovae showed that, a long time ago, the universe was actually expanding more slowly than it is today.

So the expansion of the universe has not been slowing due to gravity as it MUST be. This is very peculiar.

Additionally it was found that galaxies were not dissipating, but are still clumping together; in fact, the Milky Way is now part of a supercluster of galaxies called "Virgo" with over 100

23

of the monstrous things all affecting each other   as if something around this "cluster" is pushing things together.

The Law of Entropy is being violated by over 100 super clusters of galaxies. This can only happen is an "outside force" is overriding Entropy.

John Wheeler is the guy who named "Black holes" said the universe is built like an enormous feedback loop, a loop in which we contribute to the ongoing creation of not just the present and the future but the past as well.   This effect illustrates a key principle of quantum mechanics: Everyone has known light has a similar dual nature.

Sometimes light behaves like a compact particle, a photon; sometimes it seems to behave like a wave spread out in space.

Experiments have proven that when we observe photons they become real and if we only look for an effect, the light immediately turns to [or always was] waves of electromagnetic nothingness.   A Stanford University physicist, Andrei Linde, explained it this way.   *"The universe looks as if it existed before I started looking at it and it now looks like it existed billions of years ago.- It's necessary for somebody to look at it. You need an observer who looks at the universe. In the absence of observers, our universe is dead."* He also described a Russian fairytale about frog in sour cream that sort of explained this hard to understand characteristic. He said, *"The frogs were drowning in the cream. There was nothing solid there; they could not jump from the can. One of the frogs understood there was no hope, and he stopped beating the sour cream with his legs. He just died. He drowned in sour cream. The other one did not understand nor did he want to give up. There was absolutely no way it could change anything, but it just kept kicking and kicking and kicking [not worrying about death]. And then all of a sudden, the sour cream was churned into butter. Then the*

24

frog stood on the butter and jumped out of the can. So you look at the sour cream and you think, 'There is no way I can do anything with that.' But sometimes, unexpected things happen." This is the essence of the science of Participatory Anthropics.

One Explanation

This section is going to sound strange, but it is not stranger than finding out that 95% of the universe doesn't exist. Before we get into Anthropics and how we change our environment and our reality, let me talk about this dark matter and dark energy for a minute with respect to the soul and the spirit of a person. The soul is the part of you that can affect reality. We will get into this later but typically the soul is locked in a struggle with our carnal self, the soul can be released by our elimination of carnal thought and action. As we focus on helping others, and on things that one may call holy [reverence to the creator, focus on spiritual characteristics]. Some have proposed that what we call dark matter is not simply potential matter [Aether] but is mostly potential life [Souls]. As souls can work miracles [modifications of reality] when juxtaposed with the third part of "us" called the spirit [the light of life] free spirits would necessarily be this Dark Energy we can't see. Think of it like free thinking particles made of Aether and Gravity or free thinking Electromagnetic Energy made up of Electricity and magnetism. A freed soul/spirit combination could modify its vibration to change---reality. We need control of dark energy and dark matter and possibly have it already if we move away from carnal life with faith [the expansion of our soul/spirit powerhouse.] Sorry for the ranting, but I want you to understand that reality according to the Bible is not solidly constructed and according to science reality is not stable and this massive Dark stuff is still an unknown and what is known is just hard to understand. Possibly an example of a relativistic viewpoints will help here.

25

Relativistic Viewpoint

If you live in Florida for instance, you are spinning in a circle about 1000 miles an hour already as the Earth rotates. Because the direction is across your body, you are fatter on the equator than at the North Pole where you would go slower. If that isn't enough, we are spinning around the sun at a rate of 66,000 miles per hour. Besides that, our solar system is milling around in the Milky Way at a rate of 43,000 miles per hour. Now for the bigger numbers the Milky Way is rotating such that it takes 225 million earth years to make a galactic rotation. This means that the galaxy is turning at a rate of 483,000 miles per hour. We certainly aren't done yet as the universe appears to be expanding. While expansion is an arbitrary term, red-shift analysis tells us the Milky Way Galaxy is moving at a speed of 1.3 million miles per hour. We are moving roughly in the direction on the sky that is defined by the constellations of Leo and Virgo.

That all being said you are moving at least 2 million miles an hour so how fast is light to you? The speed of stationary light is 670 million MPH. We are going 0.3% of the speed of light just standing still. This means we are not aging the same as someone who is not zooming thorough our universe. While you may be 60 years old and weigh 200 pounds in this reality, a stationary person would only see a 59.8 year old person that was only 199.6 pounds and had a slightly redder complexion. Both people are living at the same time. If you ever stop, the younger you would no longer exist.

Let's Go Faster

Just for kicks, let's get in a rocket and go 330 million miles per hour [1/2 the speed of light] and we shine a light away from the stationary viewer that has really good eyesight. The 60 year old is now only 45 years old and the light that is shining 670 million miles faster is still only going 670 million miles an hour, but now instead of being visible light 550nm wavelength, it now is vibrating at only 225nm and it is invisible Infrared light. Now for the really bad part; you no longer have any hydrogen in your body as it has become subatomic and your carbon has changed to boron or something. If the stationary observer could see you- you would not be you, but both would be real until you stopped making the clump of boron shining an infrared light disappeared.

One thing to understand from this section is that light is a personal thing in your personal universe, while photons are shared-- light is not. The other thing is simply don't try it while you have a body.

If you don't learn anything but the above, you will be miles ahead of most people today. Relativity is only part of the instability of our reality. To look farther we need to review the science of Anthropics.

Participatory Physics

Before I go on let me apologize for all the physics stuff. They should have told you about these things in school and then you would have no issues with the Bible. I will be as brief as I can. This next tidbit is an offshoot of quantum mechanics that was developed by Dr. John Wheeler. It is called Participatory Anthropics, participatory Physics, Quantum superposition, or multiverse physics. You can imagine it is confusing if it has so many names, but almost all modern physics is based on this one theme today.

*This anthropic view of the universe has shown that **nothing exists without cognizant life being there** to "appreciate it".*

You may have heard of Schrödinger's Cat; where he presented a story about putting a cat in a box and added a type of poison. With Anthropics, he stated that the cat was neither alive nor dead until someone actually opened the box. Einstein did the same when asked if a tree falls in the woods and no one was around to hear it fall, would it make a sound. His famous answer was ----*"There is no forest and no tree."*

Second Principle of Relativity-While Anthropics was not a science at that time, Einstein's 2nd principle of Relativity stated the following. *"The second principle, on which the special theory of relativity rests, is the 'principle of constant velocity of light "in vacuo"[to an observer]. This principle asserts that light "in vacuo" always has a definite velocity of propagation.* "Without an observer light has no meaning."

Quantum Mechanics- Max Planck indicated that *all matter originates and exists as a projection of a conscious and* *subconscious creative force, which brings the particles of an atom into vibration.* This theory is the basis and it revolutionized human understanding of atomic and subatomic processes. "Every action, thought and choice we make contributes to the creation of the reality we experience as our life."

I know you are having issues with this type of science, but there are many instances where this is the only way we can understand our reality. Let me just give you a couple.

Earth in the Center-With high levels of certainty reviewing the speeds of objects moving away from the earth all the way around the planet that Earth is in the exact center. The Big Bang happened in the center as well so the earth cannot be where it is---without Anthropics. [Yes! You could say there was no Big Bang but then way are the stars still going away at such great speeds?]

Humans appear just as Carbon began being generated- As the universe ages, larger and larger elements become more and more plentiful. Initially only sub-atomics and hydrogen were here in large quantities. Anyway, to make animals and humans a massive amount of Carbon had to be available and still have large amounts of hydrogen. This means the universe is exactly the right age for humans to have evolved and all this super ancient races are really pretty much impossible. A million years from now there could be too many heavier materials for carbon based people to have been generated This universe has a reality specific for humans as cognizant observers.

Clumpy Universe- Rather than simply having Entropy work on the stars to have equal distribution between each and not enough energy to support solar systems. Our universe is full of clumped galaxies favorable for planetary existence and people.

We could go on and on, but essentially what Anthropics states is that only the NOW is 100% real, and our past and future become less and less part of this reality the farther from now that they are. Additionally, the reality we view is defined by the cognizant viewers [specifically the souls of cognizant viewers]. If everyone defines 630 nm electromagnetic waves as the color red, it is red to everyone. Here is the catch; the viewers must be as far from the debased Carnal life I briefly mentioned before that they can be to override other viewers.

Expanding Universe- As I mentioned a minute ago, it appears the universe sped up about the time humans became part of the universe as if **we,** somehow, energized it. As we move through our universe, we twist reality and never even know it.

Anthropics and Our Body- Anthropics not only describes how we control the universe, but also how we control our own bodies or the people near us. That variant of science is called Epigenetics.

Applied Epigenetics

This is the Science of DNA Activation, Healing and Repair with Vibrational medicine. Essentially it says <u>DNA cures itself</u> provided that it gets vibrated correctly. [Higher frequencies healing, lower frequencies sickness] Hopefully this is all sounding similar by now. Bruce Lipton, Ph.D. is possibly the leading Epigenetic scientist in our country. He indicated *our genes and DNA are activated and influenced by signals from outside the cell membrane. While meditation and intentions cause chemical changes in our body and brain and affect the way our genes communicate instructions to our cells and literally have the <u>effect of altering and activating our DNA</u>*.

High Frequency Consciousness Affects Reality

Concerning how epigenetics plays with Anthropics we look to biology. Rupert Sheldrake, one of the world's most innovative biologists. While many biologists try to look at the symbiotic relationship of the thousands of animals that make up a single animal or person, He revolutionized scientific thinking with his theory of *a living, <u>developing universe with its own inherent memory</u> or Morphogenetic Energy Field*. Our genes, being one of the millions of live attributes of the universe reside in this Morphic field, that controls how we sense reality, and they play an important role in the epigenetic control and activation of our DNA. This Morphic field is a field of energy containing certain frequencies of information or resonances. All living species, human beings, animals and plants, have a consciousness and therefore affect this Morphic field that exists at a particular threshold of vibration.

He indicated, *"Through conscious intention we can change our quality of life from what we have been handed down epigenetically from previous generations."* When our portion of the Morphic energy field resonates with lower vibrations associated with fear, conflict, sexual attraction and stimulation, and self- centered adoration or pride instead of LOVE. *As human beings we actually choose our experiences by the way we resonate or vibrate as an energy field. Our perceptions and thoughts translate into vibrational frequencies. Energy follows thought—shaping our reality.* Like many others, he is saying if we concentrate of Love *"love thy neighbor and we love ourselves"* we can control our environment just like Jesus told his disciples to do. As we review how all this changing of reality can be accomplished we need to look the "free" portion of our being this portion we call the soul.

The Soul of Anthropics

I know I'm losing you so let me start over. The Theory of Anthropics is not exactly centered on people. It seems to be centered on a component of people we loosely call our soul. A soul is tied to our perception of reality, but our soul is not exactly "IN" this reality. Instead, it seems to be joined together with other souls in some way and it is this union that "defines" what we perceive as reality. Let's say this union of souls decided one day to not want the color blue. Not only would there be no blue anymore. There would NEVER have been blue from the beginning of time. Pretty weird isn't it. Fortunately or unfortunately, the more we look into existence the more this oddball notion fixes the problems in definition and understanding. Einstein was not trying to be coy with his tree answer. <u>We can control our reality rather than simple observing it</u>. The more we separate from this reality, the more we control it. The various writers in the New Testament writers told us Einstein was telling the truth. Here is what Jesus told them so many times, not one forgot it.

Matthew 10:39- *He who finds his life will lose it, and he who loses his life for My sake will find it. Accept what you are able to do and what you are not able to do.*

Matthew 16:25- *For whosoever will save his life shall lose it: and whosoever will lose his life for my sake shall find it.*

Mark 8:35- *For whosoever will save his life shall lose it; but whosoever shall lose his life for my sake and the gospel's, the same shall save it.*

Luke 9:24- *For whosoever will save his life shall lose it: but whosoever will lose his life for my sake, the same shall save it. For what is a man advantaged, if he gain the whole world, and lose himself.*

Luke 17:33 *-Whosoever shall seek to save his life shall lose it; and whosoever shall lose his life shall preserve it.*

John 12:25- *He that loveth his life shall lose it; and he that hateth his life in this world shall keep it unto life eternal.*

Thomas 1:56 *- Jesus said, "Whoever has come to understand the world has found only a corpse, and whoever has found a corpse is superior to the world."*

Secret Gospel of James-*The Lord answered and said, -- Therefore, become seekers for death, like the dead who seek for life; for that which they seek is revealed to them. And what is there to trouble them? As for you, when you examine death, it will teach you election. Verily, I say unto you, none of those who fear death will be saved; for the kingdom belongs to those who put themselves to death.*

I know I was going to go through all the science before getting to the Bible. But this was such a strong theme; I thought this should be addressed right now. You may have read these things without even recognizing the significance to Einstein and Anthropic science. I'm sure many just said this is just too weird and skipped over the message that was so important during the time of Jesus that it was stated at least 7 times to insure we would **READ** IT. What they say over and over again is that *if one loves carnal life [that portion we usually see touch and feel and drives the vibrational component of our life to low levels], he will not be able to realize what life is really about [and be able to mold "reality" as needed to understand and enjoy it]. Another way to describe the message is that dying is not death but release to adapt reality.*

Let me give you an example everyone has heard about while we are still alive.

Dead DNA-I know all this sounds like craziness, but I am going to get you through the hard parts so you can understand life and with it, you will begin to understand death. Speaking of death, what is the difference between live DNA and Dead DNA?

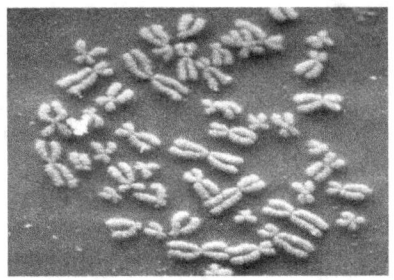 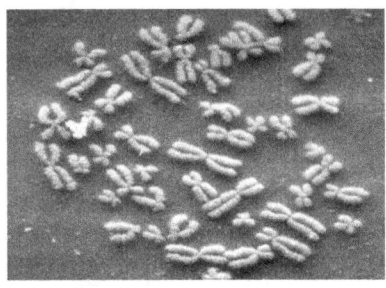

The image to the left is of the 46 DNA Strands found in a typical human cell. To the right are the same set of DNA strands after a person is "dead". What is the difference? If you give, I will tell you the answer---- "There is NO difference in the carnal existence we perceive". Dead and living DNA have the same sugars, same, interfacing bonds, same structure same chemical transfers, same biology. The difference is that a dead one is not connected to a soul. Without the soul, there is no life. There can be matter and energy, but there is no life. You can take all the lightning and energy input you can think of and the dead DAN will not become alive. That being said, many times someone is pronounced dead and after a long time he comes back to life in the hospital bed or wherever. DNA has nothing to do with it. What we know now is <u>life is from outside the body</u>. For a tiny fetus human to move around as his body reacts to this world well before the time of birth, it HAD TO HAVE OBTAINED A SOUL. Sorry, but I have to say this. The 5.5 month old fetus shown is targeted for destruction in this strange world we live in where no morals are considered acceptable. This "abortive destruction" will cause the soul to leave the body as he or she is murdered

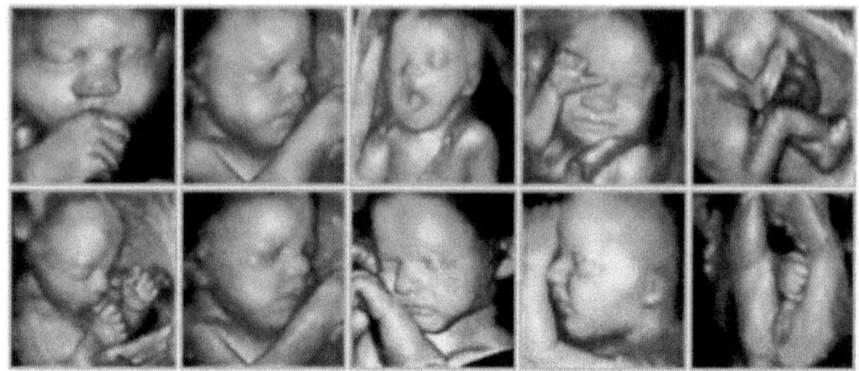

The Self-The outside of your existence is something we can call the self or the body. It is the part that is experiencing the "feelings" of existence. It is what we call carnal. One definition would be the part of us that is focused on self, survival, and sex. I know some of you are saying three cheers for self, but those tactile experiences greatly reduce someone's understanding, and interaction with the truer world. The self must be tempered with the SOUL that sort of surrounds us.

Collective Reality-Reality is made up as a collective. The color red is simply a certain vibration of electromagnetic waves, but every time you and everyone else "sees" this particular vibration, we not only build a color, <u>we create visibility</u>. I'm not saying everyone must understand RED the same way. I'm sure some perceptions of that color are different than mine, but the definition is similar enough for us to all be living in this same reality. Just think about an eyeball sensing different photonic frequencies that are phase shift ever so slightly as they bounce off things we consider to be solid. That jumbled up vibrational mess is converted into visibility almost by magic.

The Heart- Let me tell you one more thing before we move on. For years people have said their "soul" or "love" or "subconscious" came from the heart. Well guess what? We now know there is a second set of neurons [a second brain] in your body and it is located in your heart. While we still don't know a lot about this brain, we are told this brain is faster than our head brain and it reacts to feelings rather than facts and it senses endorphins faster. It seems this is a love brain. We can believe the love brain is more associated with the soul than the head brain. If our heart brain takes control of our bodies, we may think about sex, self, and survival less and there is no tell what someone might be able to do. One might be able to change reality enough to lift a car to save a loved one. I know this is probably new to you so let me give you a short description. While many are now being drawn into this new science, the forerunners are a group of 4 researchers that have been investigating this for years now. They include R. McCraty, Ph.D., M. Atkinson, D. Tomasino, B.A., and R. Trevor Bradley, Ph.D. The field is called Heart-brain Biogenetics.

Heart Brain Biogenetics

Scientists have now found that the heart is more than just a pretty face. It has its own brain, endocrine system, communication, and control over our bodies.

Heart and Emotions- This description is fairly well recognized and described in the report by *[Rein, Atkinson, et al, 1995]-It is long known that changes in emotions are accompanied by predictable changes in the heart rate, blood pressure, respiration and digestion. So, when we are aroused, the sympathetic division of the autonomic nervous system energizes us for fight or flight, and in more quiet times, the parasympathetic component cools us down. In this view, it was assumed that the autonomic nervous system and the physiological responses moved in concert with the brain's response to a given stimulus.*

Heart and Brain- The previous definition tried to stay esoteric, but this one really hits home. *[Lacey and Lacey, 1978]- However, following several years of research, it was observed that, the heart communicates with the brain in ways that significantly affect how we perceive and react to the world. It was found that, the heart seemed to have its own peculiar logic that frequently diverged from the direction of the autonomic nervous system. The heart appeared to be sending meaningful messages to the brain that it not only understood, but also obeyed.*

It wasn't until 1994 that another neurophysiologist, Dr. Armour introduced the concept of functional 'heart brain'. His work revealed that the heart has a complex intrinsic nervous system that is sufficiently sophisticated to qualify as a 'little brain' in its own right.

His report indicated the following: *The heart's brain is an intricate network of several types of neurons, neurotransmitters, proteins and support cells similar to those found in the brain proper.*

Its elaborate circuitry enables it to act independently of the cranial brain – to learn, remember, and even feel and sense.

The heart's nervous system contains around 40,000 neurons. Information from the heart; including feeling sensations, is sent to the brain through several pathways. These main nerve pathways enter the brain at the area of the medulla, and cascade up into the higher centers of the brain, where they may influence perception, decision making and other cognitive processes. Like our head brain they are masses or brain gue called Ganglia. The images following are of these ganglia and their connective synaptic nerve fibers.

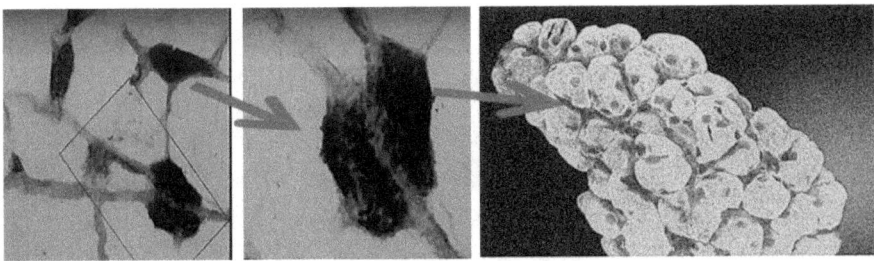

Here is a strange and possibly important part; these ganglia are distributed throughout the heart as shown next. This may give them the effect of providing faster, freer access, computation that the head brain. Each dot in the following diagram is another ganglia cluster. [Ganglia concentration positions noted with G.P.]

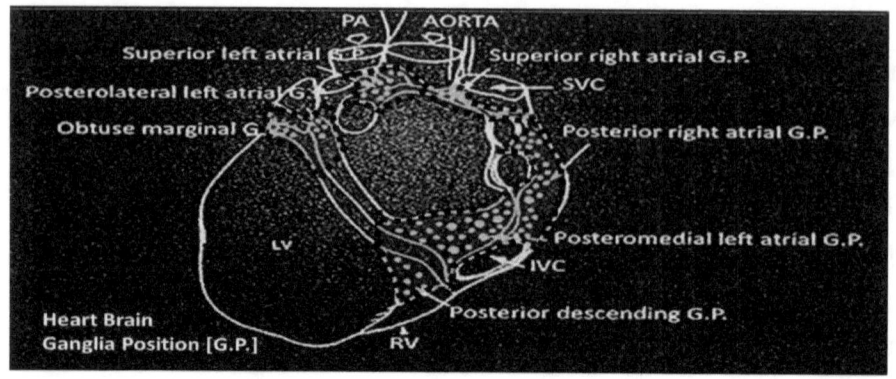

The Heart's Electro-Magnetic Field

Research has also revealed that the heart communicates information to the brain and throughout the body via electromagnetic field interactions.

The heart generates the body's most powerful and most extensive rhythmic electromagnetic field. The heart's magnetic component is about 500 times stronger than the brain's magnetic field and can be detected several feet away from the body.

It was proposed that, this heart field acts as a carrier wave for information that provides a global synchronizing signal for the entire body.

Heart Interactions Between Individuals

Additional neurophysiologists including Dr. *McCraty [2002],* jumped on the bandwagon. *He discovered a neural pathway and mechanism whereby input from the heart to the brain could inhibit or facilitate the brain's electrical activity.*

There is now evidence that a subtle yet influential electromagnetic or 'energetic' communication system operates just below our conscious awareness. Energetic interactions possibly contribute to the 'magnetic' attractions or repulsions that occur between individuals, and also affect social relationships. It was also found that one person's brain waves can synchronize to another person's heart according to work done by McCraty in 2004.

Now think about this for a minute assuming the heart as a tool for the soul. As we give our soul more allowance to "wander" this communication with others can allow us to not only affect our own reality, but also change others through this rudimentary communication link. Again when I'm talking about soul wander I'm talking about the soul [reality] in combination with the spirit [anti-carnality].

Atrial Natriuretic Factor Hormones

Besides the E-M and direct nerve interactions, the Heart commands a hormonal Army. Another component of the heart-brain communication system was provided by researchers studying the hormonal system.

The heart was reclassified as an underlineendocrine gland in 1983.

The main hormone produced and released by the heart called atrial natriuretic factor (ANF) was isolated in 1983. This hormone exerts its effect on the blood vessels, on the kidneys, the adrenal glands, and on a large number of regulatory regions in the brain. The heart allows us to get the blood we need not only by simply pumping it but also controlling how large our blood vessels expand.

Noradrenaline and Dopamine- The impressive heart was also found to contain a cell type known as 'intrinsic cardiac adrenergic'' (ICA) cells. These cells release noradrenaline and dopamine neurotransmitters, once thought to be produced only by neurons in the brain. These, together help control happiness in us so our heart helps us feel happy.

Oxytocin-More recently, it was discovered that the heart also secretes oxytocin, commonly referred to as the 'love' or bonding hormone. In addition to its functions in childbirth and lactation, recent evidence indicates that this hormone is also involved in cognition, tolerance, adaptation, complex sexual and maternal behaviors, learning social cues and the establishment of enduring pair bonds. Concentrations of oxytocin in the heart were found to be as high as those found in the brain from studies in 1986. This means ½ of all the feelings of love of your body come from the heart. Isn't that weird?

E-M Messaging to the Brain

Coherent heart rhythm patterns are sent to the brain and the effect has been recorded. The effect is often experienced as heightened mental clarity, improved decision making and increased creativity. Additionally, coherent input from the heart tends to facilitate the experience of positive feeling states. This may explain why most people associate love and other positive feelings with the heart and why many people actually feel or sense these emotions in the area of the heart. The following diagram shows the pattern of long term heart rate when a subject was frustrated and when he was calm. Notice the blood pressure ebb and flow builds a sine wave [0.1 Hz] that assures a smooth E-M Field generation while the other one sends out burst of varying frequencies that are interpreted by the brain and other organs to limit feelings of warmth, comfort, love, and compassion.

The Learning Organ

Research in the past two decades has shown that the heart is an information processing center that can learn, remember, and act independently of the cranial brain and actually connect and send signals to key brain areas such as the amygdala, thalamus, and hypothalamus, which regulate our perceptions and emotions. Once a partner is learned by the heart it seems the levels of love based hormones and the coherence heart rate frequencies come about much more often. It seems we have a second "brain" in our chest. While some of this memory might be pheromone based, it is looking more and more like Heart brains communicate with each other over their E-M communication channel. If we find out how to talk to it we may even cure lovesickness.

If you want your body to feel better and get better, being in a room full of loved ones certainly helps. If you are in a coma, loved ones may be able to communicate with your heart.

The Heart is Conscious- There is now an idea of "introspective awareness" of the heart itself. Possibly you can say we are conversing when only you, your head brain and your heart brain are in the room. While this is the scientific rage, I am more inclined to think that the soul uses the heart-brain to communicate with others. As we remove our self-centered emotion to the emotion of love and protection of others, our soul becomes very powerful [increase primary faith] according to Biblical teachings.

Heart sort of Sees the Future-I was going to tell you about another study that showed the Heart brain could understand stimuli faster than the head brain, but that is starting to sound twilight zone-ish. Oh well, I might as well tell you now. The heart can provide a split-second "body premonition." Essentially, test subjects hearts "felt" the future before it happened. Don't ask me to explain this at all as we continue. This is simply science.

Heart Replacement Caution-The heart contains a complex intrinsic nervous system comprised of multiple clusters of neurons that network with each other and acts on their own. Unbelievably, replacement of a heart is not devastating to the new owner. One would think he would not get feelings of love, have the proper hormones released and not be able to tell the head brain what it needs to, but the E-M communication network of the Heart being 500 times as strong as the E-M output of the head brain allows a stranger's heart to work in a new body with these photon like things.

Heart Controls Love-Here is a scary thought. The type of people you liked before you gained a new hear might not be as compatible with your new heart as you would like. You may have to change significant others. None of this means that the heart has a mind. It takes more than neurons, or even a system of neurons, to form a mind. A complex network of neurons can function like a computer chip, and no more has a mind than your laptop does.

More Active Than the Head Brain-According the McCraty study, *the heart is in a constant two-way dialogue with the brain and the heart system is sending far more signals to the brain than the brain is sending to the heart. At the same time the heart-brain has direct connections to organs such as the lungs and esophagus and are also indirectly connected via the spinal cord to many other organs, including the skin and arteries, so its direct neuronic messaging is substantial. Its hormonal messaging affects many body functions, its E-M communication link is 500 times as strong as the Head-Brain, it feels things for those around you, and now we believe it remembers what its likes and dislikes were.*

If you want to stay healthy you had better get a compatible mate. I don't mean someone your head-brain thinks is sexy.

Somehow we need to ask our heart who we should be with. One can make the connection that the Soul gateway to the body is through this special brain rather than the one on top of your head. Let's see what our Bible had to say.

Deuteronomy 4:29 *But if from there you seek the Lord your God, you will find him if you seek him with all your heart and with all your soul.* This verse tries to tell us about how the heart and soul are closely connected.

44

2 Corinthians 4:16-Therefore we <u>do not lose heart</u>, but though our outer man is decaying, yet our inner <u>man is being renewed</u> day by day.

Matthew 22:37 Jesus replied: '<u>Love the Lord your God with all your heart and with all your soul and</u> with all your mind.'

While this sounds like the Heart is associated with the Spirit, the third part of our being, it is really reaffirming the description of Deuteronomy.

Jeremiah 29:13 You will seek me and find me when you <u>seek me with all your heart</u>. This is firmly stating trying to understand God with only your head brain is impossible.

It was as if the Bible was way ahead of these scientists, but now we are beginning to see the heart is an important gateway to the soul and listening to your soul or heart can help us live with more love affection and more in tune with our reality. This is getting too sappy, so I'm moving back to mysterious headlines.

Lifting Super Strength

I'm thinking you are skeptical about the lifting a car thing so, before we go on, I thought I would give you a short list of well documented reports of this "impossible" type of interaction in recent times. This is only a small number of the real quantity, but it will show you how common this type of event is.

Adam Simmons was working on his daughter's Jeep Liberty when it fell on him. His 22 year old daughter, Rachael, lifted the 5,600-pound car up to free her father. He escaped with only minor cuts and bruises due to his daughter's quick action. She had no injuries and her backbone stayed in place.

In 1982 Tony Cavallo of Lawrenceville, Georgia, was repairing his Chevrolet Impala from underneath. You guessed it. The vehicle was propped up with jacks, but it fell. Cavallo's mother, Mrs. Angela Cavallo, lifted the car high enough and long enough for two neighbors to replace the jacks and pull Tony from beneath the car.

In 2006, Tom Boyle, of Tucson, Arizona, watched as a Chevrolet Camaro hit 18-year-old Kyle Holtrust. The car pinned Holtrust, still alive, underneath. Boyle lifted the Camaro off the teenager, while the driver of the car pulled the teen to safety.

In 2009, Nick Harris, of Ottawa, Kansas, lifted a Mercury sedan to help a 6-year-old girl pinned beneath.

In 2011, Danous Estenor, *of Tampa, Florida, lifted a 1990 Cadillac Seville off of a man who had been caught underneath.*

In 2012, Lauren Kornacki, *of Glen Allen, Virginia, rescued her father, Alec Kornacki, after the jack used to prop up his BMW slipped, pinning him under it. Lauren lifted the car, and saved his life.*

In 2013, in Oregon, *teenage sisters, Hanna (age 16) & Haylee (age 14) lifted a tractor to save their dad pinned underneath.*

A man named Regnier, *was driving on a Los Angeles freeway. He spotted a wrecked car on the side of the road with the driver slumped over his steering wheel. Regnier couldn't fathom leaving the man without doing something so he ripped the door off to pull the man out.*

In Ottawa, a man named Harris *lifted a car of a woman named Ashlyn. Afterwards she released from the hospital that afternoon with a concussion and some scrapes.*

In Hinsdale Illinois a 21-year-old guy *crashed his car against a guard rail and was trapped inside. The emergency responders couldn't force the door open because it would shift the young driver. The deputy fire chief then showed superhuman strength by pulling the doors off the car.*

In 2006 a woman named Lydia Angyiou *tackled a polar bear who threatened her son and two friends in northern Quebec. After the bear was shot, she was covered in blood and in shock, but somehow, all right.*

Vienna, Virginia was the sight *of another impossibility. A young woman lifted a truck to rescue her father, then saved the rest of her family from a raging fire.*

47

Eric Heffelmire, was working *under his GMC truck when the truck fell, gasoline spilled, and a fire ignited. His 19 year old daughter Charlotte lifted the vehicle and her father was saved.*

The reason I'm bringing these up is for 2 things. First, to let you understand that the stories you have heard about this type of thing being done are not flukes and, while impossible, they happened and the people doing these feasts were ordinary and they were not harmed. Don't even give me the explanation of adrenaline. What happened in these cases is that they vibrated themselves to a level that could change "normal" reality. They completely ignored themselves, so there bodies went outside the limitations of our CARNAL world. These miracles can even be done by vibrating photonic or electro-magnetic energy. The people in these stories had no idea they were changing reality. They had no idea they were creating a miracle. They simple were so concerned about someone else they forgot to think about themselves.

Sampson Super Strength

As we read in the Book of "*Judges*", Sampson had super strength that was beyond what his bones could take just as these miracles of today. He would go into a rage or whatever and all of a sudden super strength simply happened just like these other headlines. There were actually 2 guys like this described in the book of Judges.

1320 BC-Shamgar- He was one of the early Jewish Judges. Like Sampson, he was able to strike down six hundred Philistines with an oxgoad to save Israel. No one could withstand his super hero strength. Because of this feat Israel was at peace for about 10 years.

1102 BC- Sampson- We all know quite a bit about this Jewish Judge and his first wife getting burned to death. We know how he slaughtered 1000 Philistines with a jawbone, Delilah, his enslavement, and his destruction of the 3000 rulers of the

Philistines after sort of leading his people for 20 years with super strength that was seemingly impossible as his bones should have broken and his tendons should have pulled free from forces required for his feats. The thing different between these guys and the headline stories it that they seemed to be able to control their strength by something called faith, but there is confusion about what faith is so let me quickly go over how various types of faith are described in Judeo-Christian texts and how they are associated with the science of Participatory Anthropics.

Faith Definition

There are actually 3 types of faith described in Judeo-Christian literature that help us understand reality. The first could be called Self- Faith. This is when you have faith that a doctor can heal you or if you sit in a chair it will hold you up. Everyone has this type of faith, but there are 2 additional types of faith that help us understand science and religion. The second type of faith we might call "soul Faith"

Soul-Faith

Have read about the concept called "*self actualism*" that allows you to help others only after you stop concentrating on your selfish needs; or the "*power of positive thinking*" where you can gain more simply by thinking more positively; or the idea called "*think and grow rich*" where you try to convince yourself that you already have whatever you need and soon it will come to you? Every day a new book comes out that essentially tells you to quit thinking of yourself [Self, Sex, Survival] and then reality can be modified slightly to make your life better. What we find in Judeo-Christian texts is that most things called miracles are directly associated with this type of faith. What we find is that you do not have to believe in God to activate this type of faith, in fact, many oddball religions and sects have demonstrated many of the things presented in the Bible. By focusing on the Soul rather than the "self" causes the vibration level of our being to increase and allows reality to be modified. Some of the things described using this type of faith include the following:

- **Turning Water into wine or blood**- Egyptian magicians did this as did Jesus, and Moses. This changing one material into another has been done in laboratories as well.
- **Turning a stick into a snake**- Again Egyptian Magicians and Moses did this mighty deed.
- **Healing sick people**- Buddhist monks, at least 80 disciples of Jesus, Elisha and Elijah, a few individuals in the Bible not believing.
- **Awakening the soul of a dead person**- this was done by the witch of Endor in the Bible.
- **Walking on water**- Peter, Jesus as a man, and others.
- **Perform unbelievable acts of strength** – Many have lifted a car or ripped off a door to save a loved one and the Bible describes similar actions by Sampson and Shamgar
- **Moving mountains or other elements of reality**- Jesus told his disciples that even faith of a grain of mustard seed could modify reality this way.

Buddhists indicate this type of faith is the "heart" and "third Eye" chakras which allow reality to change. In our next section we will see how reality can be changed by changing it vibrational level in the lab. For this, I will introduce you to a scientist from Canada named John Hutchison.

Spirit-Faith

The hardest faith to understand is the kind that uses communication with the creator through the spirit. The spirit is the part of you that leaves the instant you die and immediately is in heaven. It is the part that the apostle Paul indicated converts our prayers into muttering that only the Creator can understand. By focusing on the Spirit and the Creator, the vibration level of our being is increased even farther

Romans 8:26-27- *For we do not know how we ought to pray, but the Spirit Himself intercedes for us with <u>groans too deep for words</u>. And He who searches our hearts knows the mind of the Spirit, because the <u>Spirit intercedes for the believers</u>.*

Ecclesiastes 7:6-7 Remember him—before the silver cord is severed, and the golden bowl is broken; before the pitcher is shattered at the spring, and the wheel broken :at the well, and the <u>dust returns to the ground it came from, and the spirit returns to God who gave it</u>. Please notice the body turns back into dust, the Spirit immediately goes to heaven and nothing happens to the soul as it stays part of this universe.

This is the highest form of faith, The Buddhist called this type of faith the "Crown Chakra" or the ability to see outside the universe while not possible without a strong connection with God through the spirit, a number of miracles are specific to this type of faith.

- **Bringing someone back to life-** This was accomplished by Elijah, Elisha, Paul Peter, and Jesus.
- **Remote healing of the sick-** Jesus was able to do this feat.
- **Bringing back a large quantity of people** back to life at the same time- This was only accomplished by Jesus as he was dying on the cross.
- **Converting a person to a new being** forgiven of sin so that he can later live in Heaven.

Unfortunately, the Biblical texts does not break down the three faith levels very well and this has led to confusion but it has not violated science. This next section tries to show you how fragile what we call reality is. What if I was to tell you scientists have levitated objects just by radiating them and have made them disappear and changed one material into another, you probably thinking about the "miracles" performed by Jesus and his disciples, but I'm talking about a Canadian.

52

The Hutchison Effect

While many scientists have experimented in the area of changing reality including John Keely in the 1880s and Ed Leedskalnin in the early 20ᵗʰ century, but John Hutchison has done some amazing "Miracles" in his home using ultra high frequency generators. The image of his head is shown next left. I didn't think about showing his heart, but you have to take my word for it that he has a heart brain as well.

  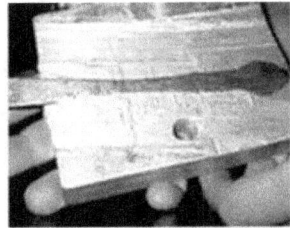

John Hutchison has demonstrated all of the following "miracles" or changes of reality in the presence of a strange field of electromagnetic waves. Keep this in mind when we talk about DNA communications.

Things passed through each other. [See the knife that floated through a piece of solid aluminum.] As they pass through each other there is no apparent change in either component physical characteristics.

Sometimes metals can become like jelly or melt without heat as shown in the middle image above.

Objects became temporarily invisible [see first two images following] and heavy objects levitated [even a bowling ball].

Let me just show you a couple of the frames of a video from one experiment in the presence of a powerful and strange field of electromagnetic waves. First pliers were picked up and yanked out of the view, but then, a bowling ball rises off the table until the ultrahigh frequency vibrations were halted and the bowling ball resumed its normal heaviness in our reality.

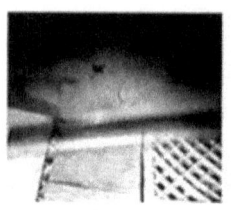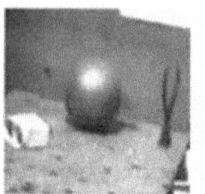

John Hutchison has accidentally found the correct vibrations for particular elements to make them appear invisible to our eyes, invisible to gravity, and invisible to each other just by changing their vibrational component. The stuff John was finding out was incredible, but someone got scared. In 2006, the Canadian Government went up to John's home and confiscated all of his vibration altering equipment. He finally got everything back, but his experiments will help us understand religion or at least some of the anomalies of the Bible. Here are some similar effects.

Disappearing and Appearing

Like Mr. Hutchison was able to do in the lab, Jesus and Philip did by simply disappearing and then reappearing.

Acts 8:26-40-Philip transported by the Spirit of the Lord out of the desert to Azotus. "And when they were come up out of the water, the Spirit of the Lord caught away Philip, that the eunuch saw him no more: and he went on his way rejoicing But Philip was found at Azotus".

Luke 4:29-31-They got up, drove Him out of the town, and led Him to the brow of the hill on which the town was built, in order to throw Him over the Cliff. Jesus passed through the crowd and went on His way. Then He went down to Capernaum--

John 8-58-60-"*Truly, truly, I tell you,*" *Jesus declared, "before Abraham was born, I am!" At this, they picked up stones to throw at Him. But Jesus hid himself and slipped away from the temple area.*

John 10:39-*At this, they tried again to seize Him, but He escaped their grasp.*

Becoming Weightless

Like Mr. Hutchison's experiments, Jesus, Peter and others made themselves have no weight and this allowed them to walk on water Here is one instance written in the Bible.

Matthew 14:25-26 *Shortly before dawn Jesus went out to them, walking on the lake. When the disciples saw him walking on the lake, they were terrified. "It's a ghost," they said, and cried out in fear.*

Matthew 14:29 *"Come," he said. Then Peter got down out of the boat ...... He said, "Come." So Peter got out of the boat and walked on the water and came to Jesus. ...*

Changing One Material into Another

Just like John Hutchison's experiments, Jesus turned water into wine. Elisha turned bad water into fresh water,

2 Kings 2:21-22- *He [Elisha] went out to the spring of water and threw salt in it and said, "Thus says the LORD, 'I have purified these waters; there shall not be from there death or unfruitfulness any longer.'" So the waters have been purified to this day, according to the word of Elisha which he spoke.*

Exodus 7:10-22- *Aaron threw his staff down in front of Pharaoh and his officials, and it became a snake.-- Egyptian magicians also did the same things by their secret arts: -- Moses and Aaron did just as the LORD had commanded. He*

55

raised his staff in the presence of Pharaoh and his officials and struck the water of the Nile, and all the water was changed into blood. ---the Egyptian magicians did the same things by their secret arts,

Here are just a few of the many other seemingly impossible things that are hinted at with Electromagnetic waves and emanations from the heart-brain.

Controlling Animals

2 Kings 2:23-24- Then he went up from there to Bethel; and as he was going up by the way, young lads came out from the city and mocked him and said to him, "Go up, you baldhead; go up, you baldhead!" When he looked behind him and saw them, he cursed them in the name of the LORD. Then two female bears came out of the woods and tore up forty-two lads of their number.

Causing and Healing Blindness

2 Kings 6:18-20 Elisha caused instant blindness of the Syrian Army at Dothan--- Later Elisha healed all the blindness simultaneously and the Army could see again at — Samaria

Multiplying Food

2 Kings 4:2-7 Elisha multiplied a widow's Oil. Jesus fed 5000 with a few fish. He did the feat a second time with 4000 in case people weren't looking.

There are many, many more, but I think bringing you up to speed on the human heart brain communications is important right about now. I think after you learn the science you are going to want to protect your heart more.

Bio Photonics

Before can understand faith healing that required the touching of someone's hand, we need to understand cells a little better. This will help us understand the Bible and its scientific exactness. It has been known for some time now that our body, all other animals, and all plants transmit, receive, and interpret photonic messages. Remember Photonic or light message and Electro-Magnetic [radio] messages are the same thing but a different frequency. Depending on how they are sent, things change in nearby cells or even nearby people or plants. While the majority of the emissions are in the Ultraviolet range, some are visible. The levels are low as the photon communications are for short distances.

The image above might show how feelings and image transfers might look if we could see them. I think the best example is when someone cuts their hand open. The body needs to repair itself, so the affected cells send out a distress "photonic message" to the nearby "good" cells" that they need to replicate. The cells get that message, replicates, and, soon the cut is completely gone. Please notice the emissions from the Heart that we talked about. The messages are by pulse coding and differences in output wavelength and its done by DNA. I'll bet you are wondering how it communicates. Well, DNA gets chemical details for the blood and it vibrates to modify light going through it. If you ever wondered why DNA is coiled like it is, this allows it to vibrate like a spring to modulate the data going to nearby cells.

DNA Emits, Detects, and Decodes Visible and UV Photons

One of the leaders in this study of light is a Russian scientist named Pjotr Garjajev. He recently was able to intercept UV Photonic communication from a DNA molecule from one organism, a frog embryo, and retransmit it to another organisms DNA, a salamander embryo, causing the latter embryo to develop into a frog! Evidently what happens is the DNA sends out a message that can be seen by cells passing nearby, if they receive a particular optical message, they might turn into skin, or whatever. The main thing is that the old idea that everything was accomplished by chemical modifications making electrical differences that were interpreted by cells has now been thrown away. Light builds people, animals, and plants.

Plants Emit, Detect and Decode

A Russian biologist was the first to find this out. His name was Alexander Gurwitsch who experimented on onion cells and found that stimulating one onion that was near another would cause the second one to flourish if there was quartz between them but not if silicon was between them because the biophotons that were being transmitted were Ultra-violet. With no barrier or quartz, one onion being fed would cause another to react. That whole concept of talking to your plants is gone------ now you need to send the right photon messages and they need to be the UV light. After the Onion Tests, I figured hospitals were not the place to be if one was sick unless people were placed in separate rooms or something that would halt UV light separated sick people. Plants would call out to those nearby with light. The image below shows a cry for help and insecticide is sprayed on a plant. The photonic emission shouts out the fear and hurt and soon settles down.

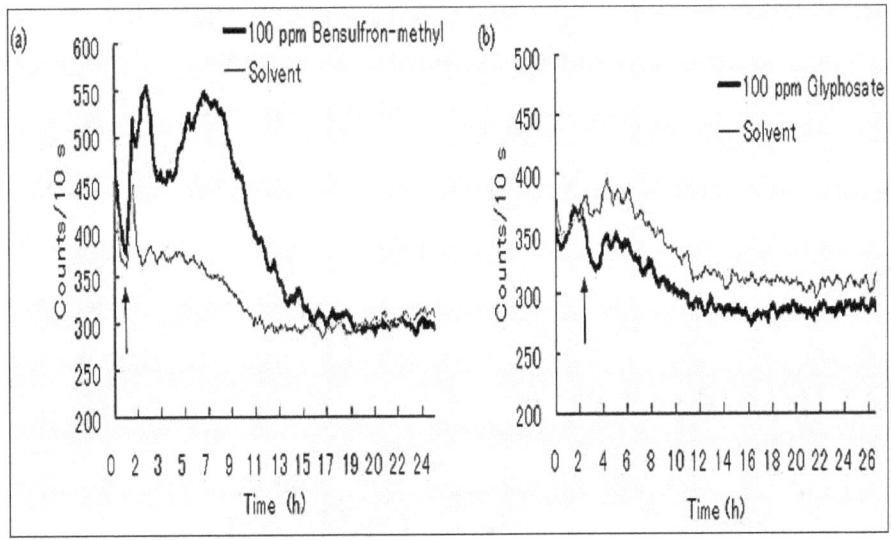

Photo-repair Sensing

Here is a strange thing! It seems that scientists found that blasting a cell with 380nm UV light so that 99 per cent of the cell, including its DNA, was destroyed was not the end of the cell. All you need to do is reblast the cell with a very low dose of 380nm and the cell will regenerate in a single day! I know that sounds like the rapid growth of cancer so let's see what happens there.

Hydrocarbons Detect and Emit

Some believe this Photo-repair messaging may be the cure for cancer. A theoretical biophysicist at the University of Marburg in Germany named Fritz-Albert Popp started his work in 1970, examining differences in a carcinogenic hydrocarbon named benzoapyrene, and almost identical but safe one named benzoepyrene. Again, UV light was the thing to activate photonic emission of these cells, He had illuminated both molecules with ultraviolet (UV) light in an attempt to find exactly what made these two almost identical molecules so different. The first one emitted a different frequency while the safe hydrocarbon reemitted the same frequency. He found out that other cancer causing hydrocarbons did the same type of photon change and they would only react to 380 nm. Once activated, a massive output of photons erupt from cancerous cells to invigorate those nearby as shown below.

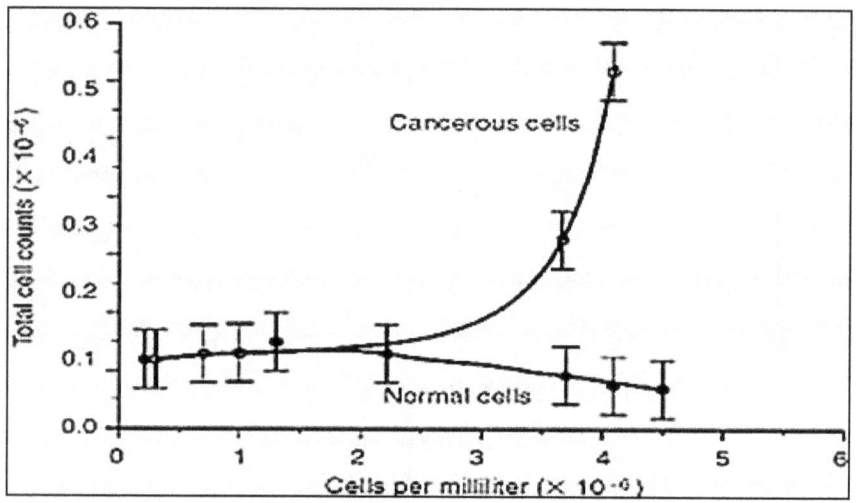

It was also noted that Melanin, in cells, is capable of transforming ultraviolet light energy into heat such that more than 99.9% of the absorbed UV radiation is transformed from potentially genotoxic (DNA-damaging) ultraviolet light into harmless heat, but destruction of that safeguard would quickly allow 380nm signals to begin changing cells.

Photons Control Everything

It seems Photons, high frequency E-M waves, switch on and control ALL the body's processes. Given different frequencies, identical cells perform different functions. The question that has forever puzzled cellular biologists for decades has been, *"What is it that enabled the tens of thousands of different kinds of molecules in the organism to recognize their specific targets?"* We now are beginning to understand how it is happening. It's not helping us define what photons are and what light is, but it is given us the details we need to begin to construct a definition.

What do Bio-Photons Do?

As Moses sort of stated in Genesis, Bio-Photons produce or establish life. They explain how enzymes can recognize their respective substrates, how antibodies in the immune system can grab onto specific foreign invaders and disarm them. By extension, that's how proteins can 'dock' with different partner proteins, or latch onto specific nucleic acids to control gene expression, or assemble into ribosomes for translating proteins, or other multi-molecular complexes that modify the genetic messages in various ways.

Seeing E-M Messages

While not getting into it in this book E-M waves enter our eyes, cause the rods and cones to active, and E-M signals are transferred, mostly but nerves, to the head brain where message decoding establishes what we call visibility, but the eyes are only a tiny part of the body that uses, transmits, and decodes E-M emissions. It seems that somehow each molecule sends out a unique electromagnetic emission that can "sense" the field of the complimentary molecule. By this, molecules recognize their particular targets and vice versa by electromagnetic resonance. In other words, the molecules send out specific frequencies of electromagnetic waves which not only enable them to 'see' each other, but also to influence each other at a distance and become drawn to each other. With about 100,000 chemical reactions happening in every cell each second and each one initiated by some special coding of bio-photonic emissions, you can see that photonic energy is switching on and off continuously. One researcher put it this way---

"We are swimming in an ocean of light."--- Just like Moses tried to say 3500 years ago.

As I mentioned, it has been suggested that the way DNA is coiled is to allow it to change its "resonance" to send and receive various frequencies needed to support life.

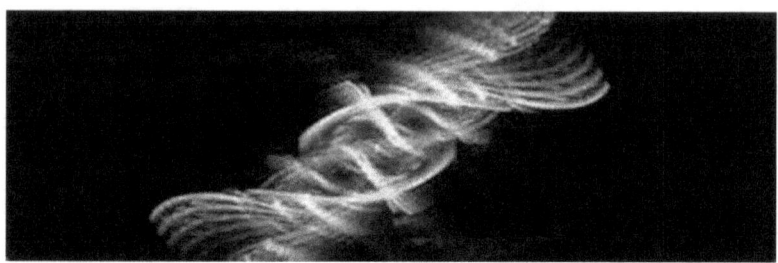

As shown above, like a tiny spring, higher frequencies would have DNA tighten while low frequency emission needs would have DNA loosen its coils to "resonate" and allow better transmission at the lower frequencies.

Here is a strange thing. It is also known that as the DNA uncoils, the amount of light emitted is increased.

We now can sense what happens when you get a cut or scratch on your skin. The images following show the extra photonic activity in areas of distress on our skin and anywhere else on or in our bodies. This is not just light. It is a message carried in the light. It is a real change in the characteristics of the skin cells. Hopefully you are comparing this data to people laying their hands on sick people and their getting better.

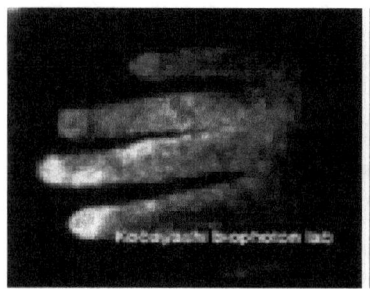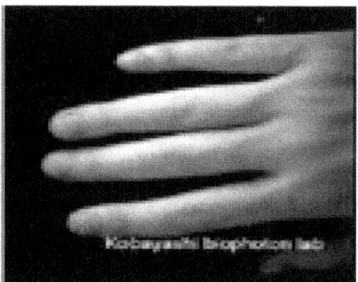

Bio-Photonic Halo

62

If you remember from Religious histories, there was mention of Bioluminescence or HALOs. Certainly, there is evidence that a concentration of bio-photonic energy could be sent through a body to help cure people, but if someone really became "charged with faith" the Bible indicated that the person would glow. The glow was most noticeable around the face as many wore clothes so it looked like their head had a Halo. We are told Noah was born with this type of radiance, and Adam reportedly had this glow. Angels covered their faces to keep people from recognizing the glow and then there was Moses. When Moses came down off the mountaintop after talking with God, he was "glowing" and had one of these halos as bio-photons were emitted in huge amounts.

*Exodus 34:29-35-Now it was so, when Moses came down from Mount Sinai--that Moses did not know that **the skin of his face shone** while he talked with Him. So when Aaron and all the children of Israel saw Moses, behold, the **skin of his face shone**, and they were afraid to come near him. And when Moses had finished speaking with them, he put a veil on his face. But whenever Moses went in before the LORD to speak with Him, he would take the veil off until he came out; and he would come out and speak to the children of Israel whatever he had been commanded. And whenever the children of Israel saw the face of Moses, that the **skin of Moses' face shone**, then Moses would put the veil on his face again, until he went in to speak with Him.*

He must have been scary and I'm certain he could have introduced a substantial amount of the correct bio-photons into someone to cure them. The image below left may give us an idea what Moses might have looked like. The following image right shows and artist's rendition. I know you are thinking he could write at night with his built in night light, but, that is just a joke.

Today, scientists are finding out this is a truth that needs to be considered as very low levels of invisible and visible light are both emitted from the body and during times of stress or disease, this output is increased. Besides the emissions, all organic life absorbs and processes light. <u>Bio-photon emission or spontaneous ultra-weak light emission has been observed from almost all living organisms, with intensities ranging from 10^{-19} to 10^{-16} W/cm^2.</u> A number of studies also found that emissions change by various cycles of the body. In Moses' case; his head turned bluish. Possibly his whole body was the same, but most was already covered. In the case of Moses, his body had a halo.

Diabetic Healing

This is very important Bio-photonic research focusing on increasing the life of diabetic patients by allowing their bodies to heal and it shows how optical messaging controls our cells. Low-energy laser irradiance at certain wavelengths is able to stimulate the tissue bio-reaction and enhance the <u>healing process</u>. This is especially important when diabetes restricts new growth so drastically. <u>Collagen deposition</u> is one of the important aspects in healing process because it can also increase the strength of the skin. It seems that biophotonic irradiation increases collagen production needed for diabetic wounds in rats. The tensile strength of skin was employed as a parameter to describe the wound. The number of cases of

64

diabetes mellitus (DM) worldwide is estimated to be around 150 million. This is predicted to *double by 2025* with the greatest number of cases in China and India. Diabetic foot and leg ulcer (DFU or DLU) is a serious complication of DM and is the single most important risk factor for lower limb amputations. More than 60% of all non-traumatic lower limb amputations are due to DFU complications. To make that even worse, around 50% of all non-traumatic amputations are as a result of DM. To further highlight the seriousness of diabetes associated lower-limb amputations, the 5-year mortality rate following amputation stands at 40 to 80%. Foot and leg ulcers are serious complications of Diabetes Mellitus (DM) and are known to be resistant to conventional treatment. They may herald severe complications if not treated wisely. That's where bio-photonics comes in. Electromagnetic radiations in the form of photons are delivered to the ulcers to stimulate healing. This first study was conducted to evaluate the efficacy of Low Level Laser Therapy (LLLT) in diabetic ulcer healing dynamics. Once the photon energy is absorbed, the photo acceptor assumes an electronically excited state. One idea is that this stimulates cellular metabolism by activating or deactivating enzymes which can alter DNA and RNA. The energy which is absorbed by the photo acceptor can be transferred to other molecules giving us observable effects at a biological level. Photon energy is absorbed by cells, which activates secondary messengers and cascades the healing optical code.

Diabetic Rats

In one important experiment, diabetic wounded rats had large wounds induced by streptozotocin via intravenous injection. I know this sounds like a mean thing to do, but once the Rat got massive wounds, the scientists could see how light or high frequency E-M waves could heal the sick. The experimental animals were treated with bio-photonic emissions at 808nm using a diode laser. The photo-stimulation effect was revealed by accelerated healing process and enhanced tensile strength of wound. Laser photo-stimulation on tensile strength in diabetic wound suggests that such therapy facilitates collagen production in diabetic wound healing which will soon reduce suffering of diabetic patients.

LED Therapy

Besides these laser tests, lower intensity RED LED emissions have successfully been used to halt and reverse effects of this horrible life stressing condition. The following image shows one of the LED emission tools used for this type of Therapy

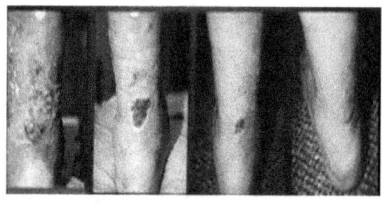

The diabetic ulsers on feet and legs are being removed by 670nm red light almost like the 808nm Laser light.

The next images show just how effective and rapid this type of theraphy is and no cutting off of legs and long hospital stays and loss of life savings. The firse image was of a diabetic patient almost ready for amputation as the leg was in terrible shape. The image to the far right above shows a typical diabetic leg ulser treated over a 4 month time with "light". The ulcers almost seem to disappear before our very eyes as cells are messaged to change from a dying state to a rejouvenation state. Its like magic or laying on of hands. [Sorry for the nasty pictures, but these are so much less nasty than many of the horrors associated with diabetes so get over complaining.]

If you are still not thinking the Biblical miraculous healings are both reasonable and scientific I feel I already have success, but there is more. Let's just go into faith healing directly.

Faith-Healing

In actual fact, in so far as faith-healing is concerned, religion is not all that important. If you remember I told you there were three levels of faith and faith healing requires only for you to ignore you self, sex, and survival to bring your body to a level that is useful for healing. There are numerous cases of faith-healers performing their faith-healing acts without using religion at all. A case in point is the science of hypnotism, the practice of which involves no religious aspects at all. Those who associate religion with faith-healing are in a way engaging in a subtle form of illusion trying to attract converts to their particular religion by making use of faith healing and describing certain cures as miraculous acts.

The methods employed by faith healers are to condition the minds of patients into having a certain mental attitude with the result that certain favorable psychological and physiological changes invariably take place. This attracts the condition of the mind, the heart, the consequent blood circulation and other related organic functions of the body, thus creating a feeling of a sense of well-being. If sickness is attributed to the condition of the mind, then the mind can certainly be properly conditioned to assist in eradicating whatever illness that may occur.

In this context, it is to be noted that the constant and regular practice of meditation can help to minimize, if not to completely eradicate, various forms of illnesses. There are many discourses in the teaching of the Buddha where it was indicated that various forms of sicknesses were eradicated through the conditioning of the mind. Thus it is worthwhile to practice meditation in order to attain mental and physical well-being. While you are reading all this remember that the cells can all be transmitting messages from the DNA and placing hands on a subject will allow for reception of the important messages.

Ruptured Disk Cure-The patient had a ruptured disk that caused him unending pain. A coworker performed hands on healing which caused a feeling of great heat. The pain left and returned 5 times before finally having ALL PAIN completely removed. After returning to the doctor it was determined that the ruptured disk had <u>miraculously</u> healed!

Lost Leg Pain-This same coworker helped another who had lost a leg in the Vietnam War. The healer prayed and laid his hands on the patient and in 5 minutes the hurting area was very hot. After one treatment, pain that had been had for 20 years was gone.

Skin Cancer-A melanoma was eliminated within one day by this same faith healer and was a visible testimony of the success of this work.

Cat Healing-No Placebo effect for sure here as a stupid cat was cured in one hands on session that was almost certainly going to die.

Traiteur Healing-It is common in Louisiana Cajun culture to have persons who have the gift of healing, thought to be a <u>blessing from God</u>, but those who practiced this ancient "ART" seemed to have been too young to understand what they were doing when they started healing. These special people are known as *traiteurs*, [Similar to the 'powow' healer of the Pennsylvania German community or the 'power doctor' in the Ozarks] They do not advertise their powers and never take money.

Buddha Healing-In Tibetan Buddhism faith healers are said to be given the "gift" from Buddha. About the same as the Christian healing, there seems to be a difference about what "faith" is. While that is not a subject of this book, let's look at the New Testament to see how it worked back then.

Scientific Description-The healings make sense given so many methods for repair data transfer.

1. DNA Photonic messaging
2. Heart Brain activation
3. Patient DNA photonic reception
4. Heart brain EM emissions
5. Heart feeling of love and affection

That brings us to the New Testament of the Bible.

New Testament Healing

Healing without Faith in God

As I mentioned before the three types of faith sometimes adds confusion, God has given us 42 clear examples in Bible texts of the exercise of this healing with ones hands. Careful study reveals a very clear pattern that definitely shows and exalts the supernatural nature of true Christian healing. Let us now look at the people who were healed. They were people from every walk of life. One of the most startling facts is that many of them did not particularly have faith in the healer. In John 9 Jesus healed a blind man whom He met in passing. The blind man *did not even know who He was.*

In verse 25 he said *he did not even know if Jesus was a sinner or not.* In verses 35 to 38 we find the man did not even know Jesus was the Christ. He did finally accept that Jesus was the son of God, but not before he had been healed. In John 4:46 the nobleman's son was never even in Jesus' presence. In John 5:5-9 the man *did not even know Jesus was going to heal him or could heal him.*

A similar instance is shown in Luke 13:12. In Acts 3:2-8 the beggar *did not have any idea what Peter was going to do.* In Acts 28:8 Publius' father lay sick of a fever so he could not have been healed by hypnotic suggestion. Many times a *man was delirious from fever* or extended sickness and still was healed.

Bad Faith Healers

By this we might understand something about faith healers that say the reason some are not cured is that THE PATIENT

<u>did not believe</u>. If you noticed from my examples, many of those healed by the "original" faith healers <u>had no "faith" at all in the healing</u>.

I think this is an important thing. Faith Healing does not require Faith in God it only need hands that do something and the ability to separate yourself from the immediate reality.

No Remote Healing

While I brought up a number of remote healings by Jesus, it should be understood that NO other person in the Biblical history was able to heal by remote intension. There were hundreds of healing by placement of hands on wounds, heads and bodies, but without very close contact, there was NO Healing. I suspect that the so-called faith healers on TV that hold their hand up in front of the camera and tell the viewers he has helped someone's horrible illness, was simply lying.

When hands are placed on bodies, there seems to be some type of energy that is transferred. If this emission is focused, it can help neutralize incorrect photonic emissions in the body of the person receiving bio-photons from the healer.

I'm not going to go over everything in the Biblical texts but hopefully the one I have provided to you have opened your eyes to understanding that Science is helped by the Judeo-Christian texts rather than making anomalies showing error or inability to use Judeo-Christian details in a world filled with science. I know modern physics is confusing and sometimes more nebulous that we would like so I would suggest reading the Bible to get the beginnings of a solution set for just about any scientific problem. One problem Einstein had was not understanding what life really was. If we are to understand death and how it is presented in the Bible, I need to pile on a little more science.

Conservation Of Energy

There was a problem with all of this. I'm not talking about the Biblical descriptions. Instead, Einstein worried more and more about all of the energy and matter leaving the universe around us until there would be no useful energy or matter universe in the area of the universe around Earth. He knew that once energy reached the end of the universe, it would be lost forever, this would include both energy associated with matter and photonic energy.

1954 Reassessment

Dimensions attached to physical shape are dashed to bits and time starts having issues as well in Einstein's Relativistic world. By 1954, Einstein was all but beat as he tried to hold on to a universe with particle defined dimensions and some magical interaction required from people. Here is what he had to say.

*According to the theory of Newton <u>the stellar universe ought to be a finite island in an infinite ocean of space</u>. This <u>conception in itself is not very satisfactory</u>. It is still less satisfactory because it leads to the result that the <u>**light emitted by the stars and also individual stars of the stellar system are perpetually passing out into an infinite space, never to return**</u>, and without ever again coming into interaction with other objects of nature. Such a finite material universe would be <u>destined to become gradually but systematically impoverished</u>.*

Ouch! Einstein realized that if the Universe is alone, soon we would have no light at all. Besides that it would have not matter and----IT WOULD HAVE NO LIFE.

"How could this be?" he wondered. In his equations it clearly showed that the universe was a sphere, but if that were so, we would be losing energy every day. Depression continued after it was noticed that Dr. Hubble's red-shifts were quantized forming a pretty important observation that will simply have to wait until we get to Participatory Anthropics or it will confuse you more than you already are. Instead, let's look at how Dr. Wolff helps us.

Milo Wolfe to the Rescue

Dr. Milo Wolfe came to the rescue during the time when it was calculated that this universe could only exist is it was linked with at least one symbiotic universe. Dr. Wolfe defined the vibrations associated with matter as "out-waves" they expanded outward to the ends of the universe. To insure that no energy was ever lost, the vibrations associated with photonic energy were defined as In-Waves as they came from "somewhere else" and continuously entered our universe. As the In- and Out-waves came in contact, they would establish something called standing waves. [We call them atoms to keep us from going crazy.] Mathematically, out-waves and in-waves looked identical except for one thing. In-waves were completely backwards. They were the inverse of matter. I suppose you already know where this is going and these in-waves were waves of energy or electromagnetic waves. Now I know some of you are remembering that "anti-matter" was really matter going backwards in time, but In-Waves are not only backwards in time, they act complexly inverted. Energy [In-Waves] and mass [Out-Waves] are the inverse of each other. Milo found that as Mass vibrations left our universe

73

and entered a linked universe they become In-Waves in that universe to establish energy and force. It was kind of like a cycle. The linked universe kept reestablishing our universe forces and matter as it leaves. This sort of can be shown as the graphic below.

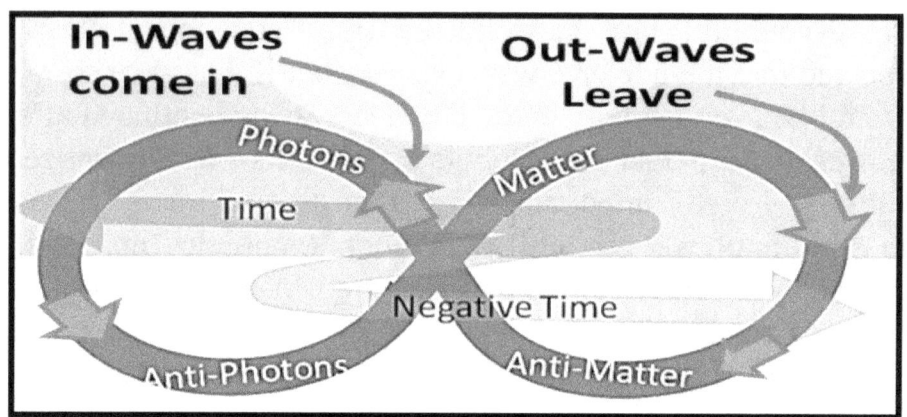

We can call this linked universe heaven as it is described in our Judeo-Christian texts. Heaven can't exist without our universe and we cannot exist without our linked universe and things are being transferred between the universes continuously. This insures conservation of everything in both universes. This is not religion this is just physics.

No Heaven---No Universe

The Bible is filled with the same thing it took Milo Wolfe to find out and Einstein fretted over for year. If the Earth got destroyed, Heaven was destroyed. The best reference is after the war of Armageddon in "Revelation". The Earth was on fire and was destroyed. All of a sudden, we read something strange.

Revelation 20-21 *When the thousand years are over, Satan will be released from his prison and will go out to deceive the nations —and to gather them for battle. In number they are like the sand on the seashore.* <u>*They marched across the breadth of the earth and surrounded the camp of God's*</u>

74

people, the city he loves. But fire came down from heaven and devoured them. --- Then I saw a great white throne and him who was seated on it. The earth and the heavens fled from his presence, and there was no place for them.---Then I saw "a new heaven and a new earth," for the first heaven and the first earth had passed away, and there was no longer any sea. [Anything happening to this world happens to the Heaven universe in what scientists call "Super-symmetry".] It's science not fantasy. Everything continuously is recycled giving the appearance of conservation of everything.

There are really only 2 laws of nature; you cannot create or destroy anything [Conservation of Matter and Energy] and everything tries to get to its lowest energy level [Law of Entropy]

Conservation

Scientists have no issue stating that "Mass cannot be created or destroyed". We also know "energy cannot be created". It can change state, but it is always here. Guess what!!! "Life cannot be created or destroyed". It can change states, but it cannot be destroyed. This is especially true in Anthropic science where the Life dimensions actually control reality much more that the vibrations of Magnetism, Electricity, Gravity and the like which only become real when an entity witnesses it. If you have ever heard the phrase that photons are sometime a particle sometimes a wave. Let me tell you something about waves----they don't really exist in what you actually understand as reality, so sometimes photons exist and sometimes they DON'T.

Now I have to bring up Quantum Mechanics, even though I hate to, but I just said photons sometimes don't exist. The quantum mechanics scientists have proven now that when a sub-particle goes into this "don't exist" state here, they miraculously appear somewhere else in the universe. They

75

cannot truly disappear. While I'm saying all of this, please understand that Einstein and all the other vibrational scientists will tell you that nothing is really here. Vibrating dimensions interacting together make waves that can be useful for these soul things to build a reality.

-

-

–

–

Please take a breath and quit getting upset with the book. Think of flowers or streams or mountains and close your eyes. While they aren't really solid, you simply don't care so you can relax. I only brought this stuff up to describe one thing.

–

-

-

-

If life cannot be destroyed, the simple answer is that there is no death. Death is an extension of life.

There are many parts to death and we will look at them, but never consider death as the end of existence. In fact, there is a cyclic nature to the life/death tango, but let me continue with science before we get into what we can call religion. The graphic following is how these guys view existence.

- Time goes forward on our side of the barrier and backwards on the other side assuring that time could be regenerated.
- Mass vibrations leave our universe to become electro-magnetic waves when considered in backwards time in the heaven universe and the opposite happens as mass leaves

heaven to be turned into electro-magnetic force in our universe

- The soul component of life abides in this universe after physical death and the spirit portion of life resides in the heaven universe as shown.

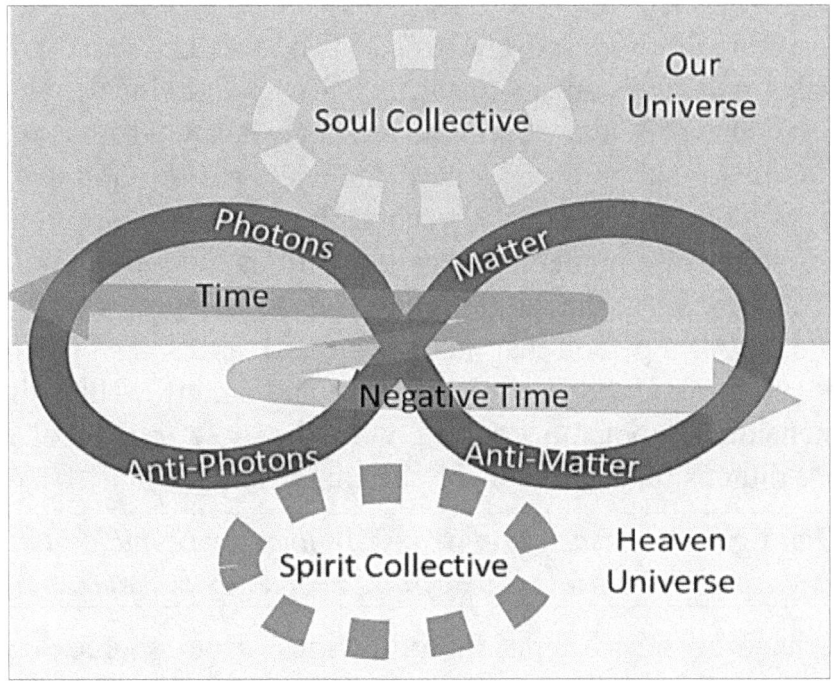

Back and forth we go always assuring time, energy and matter stay constant. Without Heaven our universe COULD NOT exist, matter would disappear, all energy would disappear, all time would disappear, and life would be gone. Luckily all this is working and we have conservation of everything and the one thing holding it all together is the collection of souls, one of the components of life itself as shown in the preceding diagram, but does the Bible make strange statements about what a person is?

What Makes a Human?

As I mentioned, the Bible indicates God is made up of three separate identities; God's word/self [The part can exist in this reality/ part that can incarnate as a living individual], God's Spirit [The part that communicates between God and man], and God's main being or soul. [This is the part of God that no one can see/ the creator/ the controller of everything.] Like God, people are made up of three identities the Self/body [the part that exists in the carnal reality], main being or soul [This is the part of a person that no one can see], and spirit [The part that communicates between God and man]. This three dimensional "dynamo" agrees with the descriptions of the other dimensional elements of our universe.

Don't get this description wrong. While we are made in his image, we are not like God beyond our makeup.

We have all heard about these "entities" from sources other than the Bible and Anthropics tells us components that make us up are <u>REQUIRED for existence</u>. In ancient Egypt these three entities were called the "Ba, Ka, and shadow". Sigmund Freud told us they were the "Ego, Superego, and the Id". Does the Bible go away from science or help explain it. Let's read a few texts.

***Genesis 2:7** And the LORD God formed man of the <u>dust of the ground</u> [Self], and breathed into his nostrils the <u>breath of life</u> [Spirit]; and man became <u>a living soul</u> [Soul].* Evidently the soul was already part of the reality God had made.

***I Kings 17:24** --And the LORD heard the voice of Elijah; and the <u>soul of the child came into him again</u>, and he revived.* The soul was just sitting outside the body and reentered it as if it were part of this universe and would not die.

Hebrews 4:12- "*For the word of God is quick, and powerful, and sharper than any two-edged sword, piercing even to the dividing asunder of soul and spirit...*" It seems here that the Soul and Spirit should always be in contact and when they are separated it is bad. Scientist knows that DNA is not alive and the Soul must activate it for life.

John 3:5-6-"*When you are reborn, you are born of spirit*" This is easy to interpret. Your soul can be reborn into a body so long as a "spirit" is sent down from the Heaven universe. We will look at this feature of life called reincarnation later.

1 Thessalonians 5:23- "*And the very God of peace sanctify you wholly; and I pray God your whole spirit and soul and body be preserved blameless unto the coming of our Lord Jesus Christ.*" All three parts of us are described over and over again.

Spirit

1 Corinthians 2:14-But a natural man does not accept the things of the Spirit of God, for they are foolishness to him; and he cannot understand them, because they are spiritually appraised. Our carnal life is focused on our self, sex, and survival. We have little time for anything else.

Job 32:7-8-"*advanced years should teach wisdom, but it is a spirit in man, And the breath of the Almighty gives them understanding.*

Proverbs 20:27-The spirit of man is the lamp/light of the LORD, Searching all the innermost parts of his being.

James 2:26- For just as the body without the spirit is dead, so also faith without works is dead. Just like sciences today, DNA cannot live alone.

Proverbs 18:14-The spirit of a man can endure his sickness, But as for a broken spirit who can bear it?

Romans 8:10-11-If Christ is in you, though the body is dead because of sin, <u>yet the spirit is alive because of righteousness</u>. But if the Spirit of Him who raised Jesus from the dead dwells in you, He who raised Christ Jesus from the dead will also give life to your mortal bodies through His Spirit who dwells in you.

Everlasting Soul

Matthew 16:26 What good will it be for someone to gain the whole world, <u>yet forfeit their soul</u>? Or what can anyone give in exchange for their soul? This is the part of people that does not die. After the body dies, the soul can live on in torment of happiness.

Psalm 103:1 --Praise the Lord, <u>my soul; all my inmost being,</u> praise his holy name.

Matthew 10:28-"<u>Do not fear those who kill the body but are unable to kill the soul</u>; but rather fear Him who is able to destroy both soul and body in hell. This is talking about fearing the Creator because the Soul doesn't die like the body.

Jeremiah 1:5-"Before I formed you in the womb I knew you, And before you were born I consecrated you; I have appointed you a prophet to the nations." Of course the Zygote from the union of a sperm and ovum is not the person, but the soul is attached to make it a living entity almost immediately. Even if we assume God waits until the Blastocyst phase a few days after conception to introduce a soul, we know it is fast. The image below shows the first few days of human development.

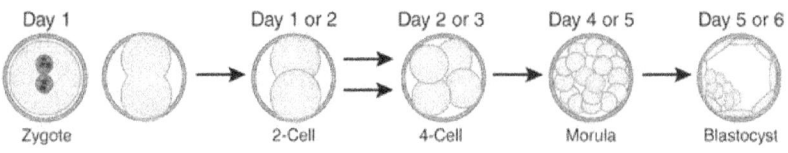

Comparison to the Other Dimensional Groups.

Electro-Magnetic Waves-First let's look at electromagnetic energy waves. We actually use these things. An Antenna radiates Electricity-Magnetism Waves that NEVER stop as shown below left. [Light is the same as it continues on to the end of the universe.]

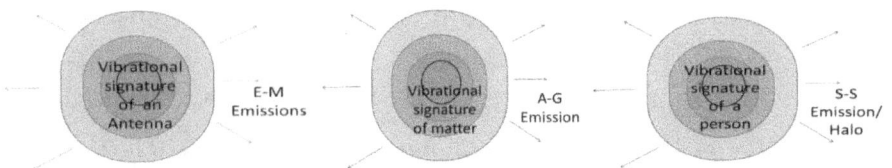

Matter-Waves- Matter does exactly the same thing. [middle diagram] While we understand matter having a surface, the Aetheric-Gravitational waves carry on without stopping [Matter has no real end.] Here is the way Einstein said it.

Since the theory of general relativity implies the representation of physical reality by a continuous field, the concept of particles or material points cannot play a fundamental part, nor can the concept of motion. The particle can only appear as a limited region in space in which the field strength or the energy density are particularly high. I know it's wordy, but he is saying matter has no real end.

Life-Waves- [See above right] When we talks about life waves we are talking about the soul-spirit waves and like everything else they do not stop at what we think of as the surface of our bodies. They continue on and on. Some have witnessed a halo around a body as the Soul-Spirit waves can be seen ever so slightly. The faster the person vibrates the more visible the halo will be. While the Judeo-Christian works call the extension of life a halo, it is well within the properties assigned by science. The portion of the halo that can be seen is the soul while the Spirit allows its existence.

81

Conservation of Time

Scientists have known for a long time that time cannot go just one direction. If time was always "advancing" soon all-time would leave the universe and there would be NONE. I know that sounds silly, but there is a simple method used in our universe to all for conservation of time, just like everything else. That secret is an adjoining time reversed universe. While some call this universe heaven, it makes no difference to me what you call it. The three dimensional particle dynamo [this is where mass and gravity are generated that I mentioned before] causes vibrational flow through the universe according to Einstein and many others. This set of waves is called out-waves in that the soon reach the end of the universe and can be lost forever if it wasn't for another dynamo of dimensions we can call the force dynamo [this is where electro-magnetics is generated]. As the vibrations from these [in-waves] coming into our universe to build stresses are generally exactly like the out-waves, except they are <u>backwards in time</u>. Guess what as our particle [out-waves] leave out universe they become "stress in "Heaven" and produce force and the Particles from outside our universe become those in-waves we need so desperately to hold masses together. Let's not look at life for a minute and just try to observe things happening in our adjacent universe. Everything would be going backwards. Things would be getting younger and newer to us as time over there is backwards.

As time leaves our universe, it is sensed as backward time in our neighbor and vice-versa. We always have time and they always have time. String scientists call all this stuff "super-symmetry". I'm not just making stuff up---you know! The diagram below, hopefully, will describe what a more up to date description that these string theorists have put together and seem to work. The universe on the left might be ours while the one on the right might be the one we call heaven. While there are billions of mass vibration centers, I only drew one at the boundary of our universe. As it travels into our neighbor, the same thing [according to super-symmetry] is coming back at us, but everything is backwards so it forms stresses [force] in our universe that we call Electro-magnetic forces that hold everything together and allow particles to build.

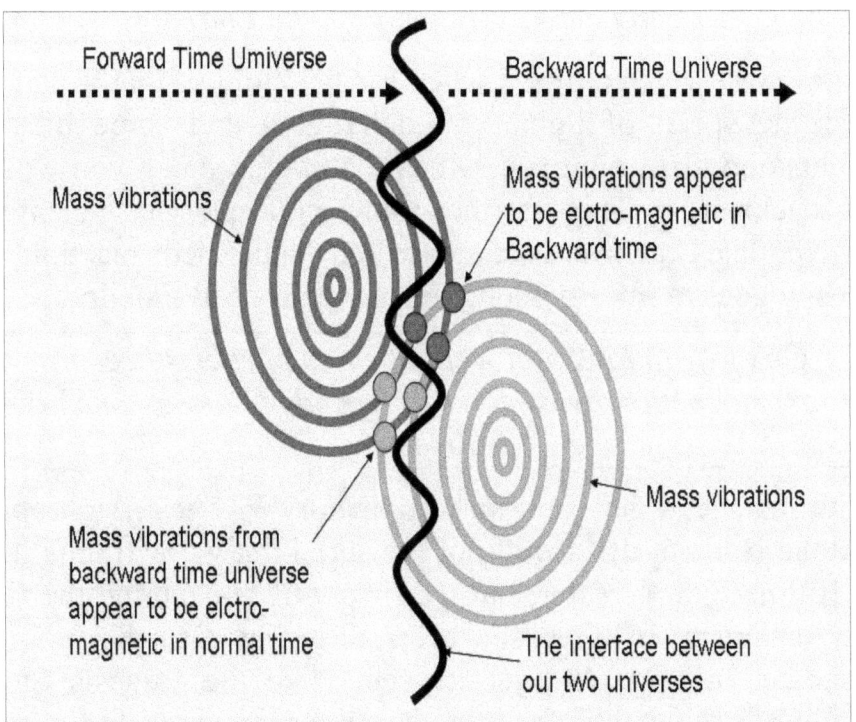

Each one of the dots drawn where the two vibration rings intersect can be considered a particle seed. From these, additional vibration rings will emerge which represent the combination of the out-waves and the in-waves.

I'm sorry for all this confusion, but scientists tell us all mass is made this way and the Bible writers tried to explain to those not understanding physics about how reality wasn't exactly real.

By the way, things would not feel backwards to people living in the adjacent universe if you were worried about them. To them, our life starts at its end and has us getting younger and younger.

Let me get bizarre on you- If a creator God was in between the Heaven and Earth universes he could see the beginning and ending of time simultaneously [We could call him the Alpha and Omega/ Beginning and End]

I know you were confused when Einstein told us that if you go near the speed of light, you quit aging and some of you understood that going faster than that would mean you would go backwards. This is the same concept, but scientists differentiate backwards time as an undefined reaction as density goes to infinity along the reduced speed timeline.

One way to look at it would be that particles become electromagnetic waves above the speed of light or when time is reversed.

With that, let's get back to Participatory Physics /Anthropics and the Biblical description of the sun. First we will look at it as if time didn't matter, but then we will re-examine the first few chapters of Genesis even allowing for the normal perceived timing. At first it seems like the Genesis Story would be problematic for scientists, but soon we realize it is so very insightful and contains science elements not well understood until our present time.

Light Before the Sun Confusion

In this section I'm going to quickly go through the first 6 chapters of Genesis as they seem to cause people problems. Certainly, Genesis 1:1-19 indicates that there were three ages with "light" before the Sun and moon were "created". How in the world can there be light without the Sun and why wasn't the sun created before everything else. Before Participatory Physics was known, the first few verses of the Bible were used against religion and to indicate the texts were not in any way factual. How could day and night happen before the Sun came along? Here is the questioned text.

Genesis 1:1-19 *In the beginning God created the heaven and the earth.--- And God said, Let there be light: and there was light.-- And the <u>evening and the morning</u> were the first day.----- And God said, Let there be <u>lights in the firmament of the heaven</u> to divide the <u>day from the night</u>; -- And let them be for lights in the firmament of the heaven to give light upon the earth: and it was so. And God <u>made two great lights</u>; the greater light to rule the day, and the lesser light to rule the night: he made the stars also. And God set them in the firmament of the heaven to give light upon the earth, And to rule over the day and over the night, and to divide the light from the darkness: and God saw that it was good. And the evening and the morning were the fourth day.*

Now that we understand a little more about reality and how it changes depending on those witnessing it, we see Moses was not sipping on fermented grape juice. He was laying out the details of a broader understanding of "nature" that was fluid. Moses was an educated man who knew that without the sun, describing light on the Earth was seemingly impossible. He did not place this in his first book to have people laugh at him but to expand the reader's awareness of the flexibility of reality. Just because the Earth is in the center of the universe did not stop the Big Bang from happening in the center of the universe. And light is even less of a challenge in that electromagnetic waves, we call light are all around us, all one must do to experience light is to shift the vibrational component, or describe light as any electro-magnetic emission. Without eyes and a method to interpret phase shifted signals entering light is whatever you want it to be.

If all of this the now being the only reality and the past and future are both built around it stuff is confusing, don't worry. This is simply one way of interpreting the first part of the Genesis story. It also supports scientific reason in linear time [The past coming before the now] but we do have to understand that Hebrew words were not plentiful so a single word meant many things. One such multi-word is Yawm or "Day".

Yawm/Day

The other thing that should be noted is that three day past before the Sun came along for the earth to rotate in front of make us have a day and night. That being said, let's simply state that this first day did not last 24 hours. This debate about timing and the Genesis story is inappropriate. Let's see what the Bible says about time. Genesis indicates it took 6 "yowm" to create humans. This has been interpreted as 6 days, but "days" was not what the Bible says. Let me give you a few examples of how the word yowm is "interpreted" into English, because the word actually means "age or time" in a general sense unless attached to other modifiers. People insist on taking the word literally, unless they decide not to take it literally. Here are some of the "I don't want to take it literally" instances.

Numbers 25:18 - *the day of the plague.* Actually, the plague lasted more than a day so yowm didn't mean day but it can be stated as day just the same.

Numbers 28:26 -*the day of the first fruits.* Actually, the first fruits last more than a day so, again, yowm didn't mean day.

Deuteronomy 20:19 *When thou shalt besiege a city a long day.* Laying siege on a city one day, even a long one, would not work so yowm didn't mean day.

Deuteronomy 22:7 -*thou mayest prolong thy day.* Most people live longer than a day so yowm didn't mean day.

Joshua 3:15 *Jordan overfloweth all his banks all the day of harvest;* It takes longer than a day to harvest so yowm didn't mean day.

2 Samuel 23:20 -*slew a lion in the midst of a pit in the day of snow:* One day of snow is unlikely so yowm didn't mean day.

1 Kings 11:42 *And the day that Solomon reigned in Jerusalem was forty years.* A day and 40 years can't be the same so yowm didn't mean day.

1 Kings 14:20 *And the day which Jeroboam reigned was two and twenty years:* 22 years is not a single day either so yowm didn't mean day.

2 Kings 3:6 *And king Jehoram went out of Samaria the same day, and numbered all Israel.* An extremely fast census could not be accomplished in a "Normal day".

Job 15:23 *he knoweth that the day of darkness is ready.* Again, they were not speaking about a single day.

Job 30:16 *the day of affliction has taken hold upon me.* Job was afflicted more than one day.

Psalm 41:1 *the LORD will deliver him in the day of trouble.* I cannot believe that there is only one day of trouble.

I could go on and on but I think you have got the picture. The Genesis writer is only categorizing ages and not specifying a specific time period, so don't get caught up in that debate.

Another Genesis Construct

Another equally consistent with science reading of the first book of the Bible In the very first part of the first chapter of the book of Genesis, we initially read the following:

Genesis 1:1-5-*In the beginning God created the heavens and the earth. The earth [became] formless and empty; darkness was over the surface of the deep. Darkness was over the surface of the deep. God said, "Let there be light," and there was light. God saw that the light was good. He separated the light from the darkness. The evening and the morning were the first day.*

Here are some issues that may help out with this whole Bio-photonics stuff.

- It says *"God created the Heavens"*, but he didn't create the sun, moon, and stars yet, so there is question about what the heavens were.

- The word *"became"* shows that the earth and heavens originally had form, but lost it.

- *Darkness was over the surface of the deep* is also strange in that God had not made the sea yet. All of a sudden *God was hovering over the waters* when there was not sea.

- This making of *"light"* was "four days" or 4-periods before he created the Sun, stars, and moon, that would

88

eventually produce light, so the light created in this section came from something else.

- Finally, God separated this "light" from the *"Darkness of the deep"*.

To read this little section one must understand a few things besides Yawm just meaning "age" or "Time period".

- The word waters/*Mayim* actually means "life giving juice or simply life giving place as water could give life". If the word was to be regular water it would have been *Ahava*, which means "regular water".

- The word Darkness/*Chashekh* means misery or destruction type of darkness.

- The word deep/*Thom* means the abyss type of deep.

Given a more modern way of saying this section without changing the words, we can see the following is easier to understand.

Interpreted Genesis 1:1-5 In the beginning God created the heaven universe and the universe holding the earth together before this first period. Because of War and Extinction, the earth was completely destroyed and so was the rest of the abyss of our universe. While God was viewing the death and destruction he remade the "essence of life" and he separated life and light from darkness and death. This all happened during the first period of the new beginning.

2nd and 3rd Age- As the ages past, Moses described the asteroid belt between the habitable planets and the Jovian ones, the re-distribution of land and the recreation of trees and grass before making light perceptible. Then came the 4th age.

Genesis 1:14-19-And God said, Let there be lights in the firmament and let them give light upon the earth. God made two great lights; the greater light to rule the day, and the lesser light to rule the night: he made the stars also. And God set them in the firmament of the heaven to give light upon the earth, And to rule over the day and over the night, and to divide the light from the darkness: and God saw that it was good. And the evening and the morning were the fourth day.

Here we need to define what light is. Today RF emissions are not usually considered light but let's see what happens. At slightly faster frequencies these RF emissions become heat. Faster still they become visible, but with faster vibrations the light disappears again into ultraviolet, X-Rays and cosmic energy without changing a thing except perception. A possible modern writing of this same section might be the following.

Interpreted Genesis 1:14-19-And God created the perception of light in the sky to divide the day from the night. God made light from the Sun and moon to be perceived. He also made the light from stars to be perceived and to divide daylight from darkness: and the evening and the morning were the fourth Age.

5th Age- God began re-creating a wide assortment of animals in the water and birds in the air that were lost during the massive Heaven war and worldwide extinction. God was now ready to provide the Anak people with a helping human to replenish the earth beyond the Anak people. We find out about the first humans and the Anak in Chapter 6.

*Genesis 1:24-2:4 And God said, Let the earth bring forth the living creature after his kind, cattle, and creeping thing, and beast of the earth after his kind: and it was so.- And God said, Let **us make man** in our image, after our likeness: and let them have dominion over all the earth. - And God blessed them, and God said unto them, - REPLENISH the earth, -And God said, Behold, I have given you every herb bearing seed, - and the fruit of a tree; to you it shall be for meat and to everything that creepeth upon the earth, I have given every* *green herb for meat* -and the evening and the morning were the sixth Age. Thus the heavens and the earth were finished and on the seventh Age God rested from all his work.

This one is pretty self-explanatory as animals and man were recreated and no animal or this particular man were to eat meat during the 6th and 7th Age as God temporarily halts "creation" here and begins a season of watching his re-creation for a time. It is believed the 1st through 7th Age were what we call the Tertiary period of Earth's development as the Earth had to be rebuilt after the Cretaceous destruction and extinction. While Moses does not talk about the Anak people that lived during this times in chapter 1; chapter 6th provides backup information to show that were here the whole time.

Interpreted Genesis 1:24-31 And God said, Let the earth bring forth the living creatures through procreation.- And God said, Let us make man in our image, after our likeness: and let them have dominion over the lower animals. - And God blessed this new man and said, procreate, and replenish the earth after the de3struction of the war and extinction. -And God said, Behold, I have given you every herb -and fruit of a tree; to you it shall be for food-- and to all the lower animals I have given every green herb for food –This ended the sixth Age. God sat back and watched how everything developed during the 7the Age and blessed this time.

Then Moses switches to a new term for "God" he now calls him "Lord God" to show how special his next creations were and we are staring on the 8th Age. This if believed to be the beginning of the Pleistocene Age [Sometime between 20 to 40 thousand years ago].

Genesis 2:5-9These are the generations of the heavens and of the earth when they were created, in the day that the LORD God made the earth and the heavens, And every plant of the field before it was in the earth, and every herb of the field

91

*before it grew. And the **LORD** God formed man of the dust of the ground, and breathed into his nostrils the breath of life; and man became a living soul. And the LORD God planted a garden eastward in Eden; and there he put the man whom he had formed. And out of the ground made the LORD God to grow every tree that is good for food; the tree of life also in the midst of the garden, and the tree of knowledge of good and evil.*

While Adam was made using elements of the universe, his DNA was not alive until God breathed life [a spirit] into his body. We also notice that while the 6th Age human talked about a group "helping" God make man, this new one had no such help as Adam would be made without the Anak interference. Here we find three components of man [Carnal body from the Earth, Life spirit from God together became a living soul].

Interpreted Genesis 2:5-9 After the Lord God observed his creation for a time, he formed man a new man from the Earth, breathed in a spirt ,so that he became a living soul. And he made a garden around him for beauty and food. The tree of life that was used by this new man and the Anak to stay young was there, and the tree of knowledge of good and evil that was not to be used by this new man was also there.

Trouble in paradise- Soon Satan, the leader of the Anak people, got Adam kicked out of the Garden. In this case Conjurer or Serpent are the same Hebrew word and yes Moses was not saying snakes had legs and a huge brain to make them able to be cunning. He never intended that people would suggest they were so handsome that they could seduce a woman. Moses used the Serpent image from Egyptian mythology as Apep, the Serpent god, was god of the sun and very powerful as was Satan. The first use of the name Satan is found in *1 Chronicles 21:1 during the time of Ezra, a*

92

thousand years after the time of Moses. Right now he was the Serpent god. While Eve had been seduced by Satan, Adam must also have had sex with the Anak people as both make aprons to cover their genitals in shame.

Genesis 3:1-13 Now the Magician/conjurer was more subtle than any beast of the field and he said unto the woman, if you eat from the fruit of the Tree of Knowledge - ye shall be as gods, knowing good and evil. And when the woman saw that the tree was good for food, she did eat, and gave also unto her husband with her-And the eyes of them both were opened, and they knew that they were naked; and they sewed fig leaves together, and made themselves aprons. --And Lord God said, Hast thou eaten of the tree, whereof I commanded thee that thou shouldest not eat? And the man said, "The woman gave me of the tree, and I did eat". And the woman said, "The conjurer seduced me, and I did eat."

With the suggested updates, the verse looks like this.

Interpreted Genesis 3:1-13- Now the Satan was more subtle than any beast of the field and he said unto the woman, if you eat from the fruit of the Tree of Knowledge you will be as the Anak people, knowing good and evil. After being seduced, she did eat, and gave fruit to her husband who had also been seduced and they became ashamed so they hid their genitals with fig leaves. --And Lord God said, Hast thou eaten of the tree of knowledge? The woman said, "Satan seduced me, and I did eat". Adam said, "I was seduced and also ate!"

Punishment- The conjurer was forced to be on the Earth forever, he no longer was sexually appealing nor would his seed produce offspring. This would bruise his ego such that he would continually try to harm the new man. The woman's punishment included having great sorrow concerning her children's lives of hardship, and she would no longer desire the Anak and love only Adam. Adam's punishment was that his easy farming days were gone.

Genesis 3:14-19 And the LORD *God said unto the conjurer, because thou hast done this, upon thy belly shalt thou go, and dust shalt thou eat all the days of thy life: And I will put enmity between thee and the woman, and between thy seed and her seed; it shall bruise thy head, and thou shalt bruise his heel. Unto the woman he said, I will greatly multiply thy sorrow and thy conception; in sorrow thou shalt bring forth children; and thy desire shall be to thy husband, and he shall rule over thee. And unto Adam he said, Because thou hast eaten of the tree, cursed is the ground; - In the sweat of thy face shalt thou eat bread.*

Today, Moses would have stated it this way.

Interpreted Genesis 3:14-19 And the LORD *God said unto Satan, because you have done this, you will be lower in my eyes and be forced to stay on the earth you whole life. You will never be sexually desirable to my new humans and you will not be able to procreate with them. Unto the woman he said, I will greatly multiply you sorrow by what your children will have to endure; and you will no longer have sexual desire for the Anak. And unto Adam he said, because thou have done this you will have to work for your food. .*

Description of the First People and the Anak- Moses backs up and indicates the ancient people were giants, The offspring of the Anak with "normal people were also giants and they all were wicked and did horrible evils. One of the things it says is that the animals and birds had become corrupted and he would destroy them along with the people. It indicates the destruction would be 120 years after the time of this description.

Genesis 6:1-7- There were giants in the earth in the early days; and also after that, when the sons of God [Anak people]

94

came in unto the daughters of men, and they bare children to them, the same became mighty men which were of old, men of renown. And God saw that the wickedness of man was great in the earth, and that every imagination of the thoughts of his heart was only evil continually. And it repented the LORD *that he had made man on the earth. And the* LORD *said, I will destroy man whom I have created from the face of the earth; both man, and beast, and the creeping thing, and the fowls of the air; for it repenteth me that I have made them. ---And the* LORD *said, My spirit shall not always strive with man, for that he also is flesh: yet his days shall be an hundred and twenty years [before the flood comes].*

In a more modern style, this section could be described this way.

Genesis 6:1-7- *There were giants in the earth in the early days and* after the war, *the Anak people had sex with the female offspring of Adam and they had children that were also giants and men of valor. During this time, God saw that the wickedness of man was over the whole earth in war and mis-creating of animals. God said, I will destroy man from the face of the earth; both man, and mutated animals, for they are no longer worshipping me. ---And the* LORD *said, I will not cause their destruction for another hundred and twenty years.*

Speaking of giants on the Earth before Adam, let's look at the Physical evidence.

First Humans of the Bible

Genesis tells us the humans created during the 6th Age were to REPLENISH the population of people who had died as a result of the extinction described in Genesis 1:27-31. How could this be? Evolution tells us only the dinosaurs existed in the Cretaceous period before the great extinction described as the time the Earth became without form and void. Some have used this confusion to indicate the Bible cannot be telling the truth.

The Greek called the first "giant" people Titans, but Moses simply called them *"giants on the earth in those days"*. *Here are some more Judeo-Christian texts that give us more details.*

Creation Text [Gnostic Jewish text]-*Now* **the first Adam** *[ancient human] is* **spirit** *endowed and appeared on the first day.* This first man lived during the Jurassic and Cretaceous periods. As this says these "spirit endowed" humans became Angels as their bodies died and then they became the Anak after the war.

The second Adam *[primitive man] is* **soul**-*endowed and appeared on the sixth day."* The second or 6th Age man would have lived in the Tertiary period. Modern scientists classify him as Homo-Erectus man.

The third Adam*, [the true Adam] is a creature of the Earth that is the man of the law. He appeared on the eighth day and became numerous and produced every kind of scientific information of the soul endowed Adam."* This would be the human created at the beginning of the Pleistocene Age- called the 8th Age in Genesis.

All three major creations of humans were laid out here so there would be less confusion. The first "giant" humans became spirits or angels which makes sense as there is no record of angels being created [They are different than Archangels who were created during the first Age.] This description is repeated in a number of Judeo-Christian texts and generally confirmed by something we call the Heaven War. Here is another texts describing the transformation of the first humans to become angels, with many returning to be the Anak people.

Codex Junius II [6ᵗʰ century Anglo-Saxon Old Testament] *Ll:19-33- He first created Adam, and a noble race, the **angel princes**, which later perished utterly. For, it seemed to them in their hearts it well might be that **they themselves were lords of heaven**----- they must needs endure grim woe and surging flame, no more possessing radiance of glory or high-built halls in heaven; but they must needs plunge downward to those depths of fiery flame, down to the bottomless abyss, insatiate and rapacious. God only knoweth how **He hath condemned that guilty host**.* While this doesn't speak about why the first humans who became angels were punished horribly, but other texts tell us there was a massive war in heaven. Because heaven and earth are linked, both universes contained massive destruction as the Earth became void and without form.

Physical Evidence of the First Humans

Anyone trying to find evidence of these ancient people do not have to look long. From substantial artifacts; bones of giant modern looking people around the world; remains of electrical devices; evidence of "growing" bricks together; extremely ancient manufactured goods; many footprints walking on the same beaches as dinosaurs; a shoe crushing Trilobites; and a Mesozoic aged group of 16 different Uranium processing plants in Africa; we can make the "assumption" that the people described by Moses were not only here and real, they were very civilized and lived for many thousands of years. The expanded texts of the Old Testament also help us help others understand how people could have walked with dinosaurs as hundreds of footprints of both human and dinosaur walkers are being found everywhere as shown next.

They also help us explain why modern looking shoeprints can be found encased in stone. Massive shoe and footprints in Australia are shown next, but that is certainly not all.

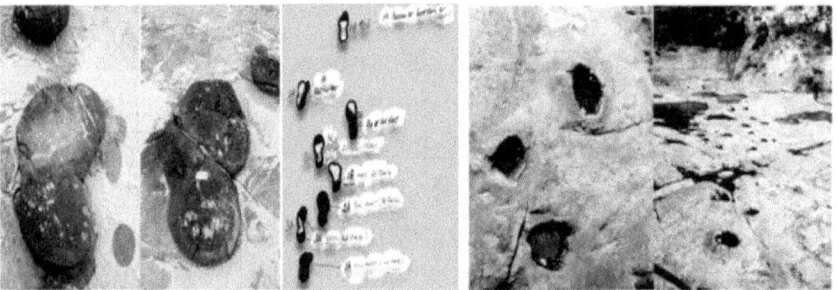

Finds from Scotland, Virginia, Washington, Utah, and other parts of the world are shown below. The Utah guy stomped trilobites according to the shoeprint. [Third from the left bottom row]

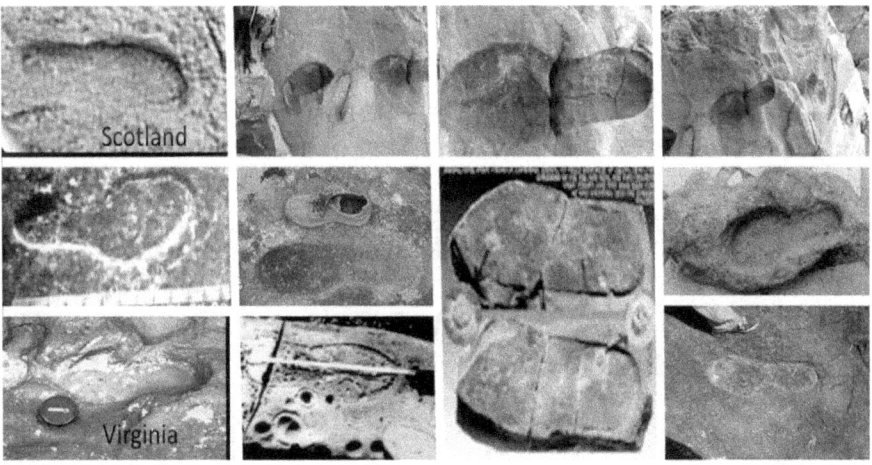

They also help us explain how people could be smart enough to build batteries during ancient times. The picture following [Left] is some type of power conversion device found <u>inside</u> a geode, found in California. Below the geode is a drawing of x-rays of the geode showing the elemental parts. These include a spring, core, plate, and electrical insulator. The same parts as you would expect in a battery. Maybe this is a new way to package batteries, but it takes a long time to complete the package. The central metal core surrounded by the white material looks like a battery. Whatever it was, it was electrical. On the right is a drawing of the parts and a size comparison to a standard D-cell battery.

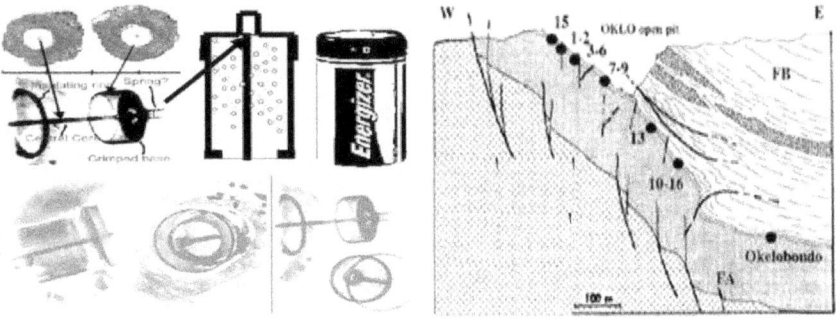

Some even have a difficult time with the extremely ancient 16 nuclear processing plants found in Oklo, Africa as shown previous right. They also can hardly believe that ancient people could grow blocks even with examples being found all over the world today. These texts can help us understand what the people of Jesus' day knew from their expanded Old Testament. The examples below are from West Virginia and Ohio, but we can find similar "grown" blocks in Peru and Australia.

As these people died, they became what we call watchers or angels who simply changed universes to live in Heaven.

Heaven Confirmation

Of course the descriptions of Heaven are provided throughout Judeo-Christian literature, but let me just start off with the beginning of the Old Testament.

Genesis 1:1-2 In the beginning God created the heaven and the earth. -- And God said, Let there be a firmament in the midst of the waters [Inhabited Areas], *and let it divide the waters from the waters. And God made the firmament, and divided the waters which were under the firmament from the waters which were above the firmament: and it was so. And God called the firmament Heaven.*

While there was some type of separation between this Heaven place and the Earth place, they were only separated by a firmament as if they were placed on top of one another. What we find is that Heaven has a similarity to our universe with at least 7 different "heavens" and maybe many more. These could be either 7 different universes [All linked to our universe, or they may simply be part of a single universe with planets or something similar] Except for the second and fifth, most of the heavens described are all wonderful places, according to Enoch and the Bible. The "New King James or Baskerville version of the Bible [1827]" talks about the wonderfulness of the "third heaven"

II Corinthians 12:2- I know a man in Christ who fourteen years ago was caught up to the third heaven. Whether it was in the body or out of the body I do not know—God knows. And I know that this man—whether in the body or apart from the body I do not know, but God knows— was caught up to paradise and heard inexpressible things, things that no one is permitted to tell.

In the early King James Version [1611 to 1827] the 5th heaven is described. *III Baruch 11:2- "- and the angel took me and led me thence to the 5th heaven- and I saw the commander Michael –"*

Gnostic Jewish Texts tell us more

II Enoch 3 through 21 gives us additional details and we see that the numbering is arbitrary. Baruch's 5th heaven is Enoch's 4th heaven. Next is a table describing what Enoch saw.

Heaven #1	*Called Saamayim, this is just above the sky. Angels abide there so that they can have quick access to help people in need and protect them. The angel Gabriel rules it.*
Heaven #2	*Called Raqia, this is a place of darkness, full of prisoners weeping continuously [devil's angels]. The angel Zahariel rules it.*
Heaven #3	*Called Shehaqim, this is where the tree of life and the garden now reside with 300 angels to tend and secure it. The angel Anahel rules it.*
Heaven #4	*Called Machonon, this is where [12 winged ones] and 6 winged angels sing continuously. The angel Michael rules it.*
Heaven #5	*Called Mathey, this is where the devil is chained. The angel Shadalphon rules it. [By the way, the devil and Satan are not the same being—nor is Lucifer for that matter. The proof will be provided.]*
Heaven #6	*Called Zebul, this is where the archangels reside over the regular angels. Zachiel rules.*
Heaven #7	*Called Aroboth, this is where the troops of God reside including cherubim, seraphim, and Ioanit who rally and sing to God. The angel Cassiel rules it.*
Heaven #8	*Called Muzaloth, this is where the stars are formed.*
Heaven #9	*Called Kuchavim, this is where the heavenly ones reside [Probably more archangels as we discussed in book one.]*
Heaven #10	*Called Arovoth or Parvain, this is the house of the Lord God, creator.*

Creation Text- "Now the prime parent [God] created heavens for each of his offspring through verbal expression. He created them beautiful, as dwelling places, and in each heaven great glories times seven. Each one has these in his heaven; mighty gods and lords and angels and archangels."

Apocalypses Moses- *"See with thine eyes the <u>seven heavens</u> opened and see thy father lies on his face and all the holy angels are praying -Paradise is on the <u>third heaven</u>."*

Testament of Levi 1: 10-27-"The 1ˢᵗ heaven had a <u>hanging sea</u>, the 2ⁿᵈ heaven had <u>boundless light</u>, on another was the <u>host of angels</u>, another had the <u>holy ones</u> [archangels], another held angels that answered to the lord, the next had the thrones and dominions, and the highest held the Great Glory."

With the various "heavens" in Heaven having names it seems more likely that these are regions in a single universe, but who knows. The main thing is that scientists have not only confirmed a linked universe but also tell us it is a requirement for us to even exist. By the way, the Bible indicates when the Earth id destroyed at the end of days, Heaven is destroyed as well and a new heaven and new earth must be made showing our two universes are symbiotic.

Revelation 21:1-Then I saw "a new heaven and a new earth," for the first heaven and the first earth had passed away, and there was no longer any sea.

This passing away not only happens at the "End of Days", it also happened at the end of the Cretaceous Period.

"Heaven" War Confusion

Our Bible is filled with references to a horrible war between angels who were tired of the restrictions in heaven and those who loved it there. The first question some bring up is "How could angels not like heaven unless that had experienced carnal, sex filled lives before becoming angels who had a more reverencing life?" If you remember these angels had once been the giants who walked with dinosaurs so the main reason for the war was a desire for "human-type living with all its carnal pleasures. Sure, Heaven was beautiful beyond belief, happiness down to your toes was experienced, one could be close to the maker of the universe, and on and on, but some of the angels wanted sex. They wanted to get drunk again, they wanted to go into battle, and they wanted to change Heaven to be more like our universe had been. From this first experience, we know that God is much more particular who he allows in heaven today as he has no desire for a repeated war. Let's read just a few of the many texts about this war.

Genesis 1:1-2 In the beginning God created the heaven and the earth. And [then]the earth was without form, and void; and darkness was upon the face of the deep.

Jeremiah 4:23-24-[near the end of the wars] "I beheld the Earth, and, -- all the cities thereof were broken down. If the cities were broken down, there must have been cities before the war.

Isaiah 9:17-21---`Is this the man [Satan] -- made the world like a desert and overthrew its cities, All the kings of the nations lie in glory, each in his own tomb. Kings of nations and Cities shows a substantial civilization before the world became like a desert.

Isaiah *13-1 They [Heavenly Host] come from a far country, from the end of heaven, even the LORD, and the weapons of his indignation, to destroy the whole land* [and make it without form and void].

Here is something to remember about the universe of Heaven while it is not exactly here, it may very well occupy the same space as our universe but we simply have no "understanding of it in our reality". Certainly it could also be billions of mile away, but we know Heaven is a requirement for our universe to survive. We only get a glimpse of this war in the Biblical statements I brought up so far, but maybe we can get a little better picture. Here are a few more tests.

Codex Junius II [6th century Anglo-Saxon Old Testament]- *Within God's kingdom in the days of old, the angel prince was called "Light-bearer", Satan. But he stirred up strife in heaven and turned to insolence and pride. Darkly, Satan planned to build a lofty throne in heaven.*

Creation Text [10th century Jewish Gnostic]- *and Samael [Satan] said," I have no need for anyone, It is I who am God, and there is no other one that exists from me." God was filled with anger and said, "You are mistaken, Samael, there is an immortal man of light that has been in existence before you [God], and who will appear amid the creatures you have made, and will trample you, and you will descend to the abyss. --- Then he and his followers made a great war in the seven heavens.*

When a text talks about creatures being made, they are talking about genetic manipulation, not creation. "Enoch II" provides

105

a good description of the 7 Heavens, if you are interested, but I want to try to stay on earth as much as I can so we can amplify the Bible for our understanding. When Paul saw the 3rd Heaven after temporarily dying, he, like Enoch, reveled at its beauty.

Midrash 13- [2nd **century Jewish Cabala Text**]-*Samael [Satan] took his cohorts went down and saw all the creatures that the holy one had made. The world had just been created, is this the time to <u>rebel against God</u>?*

Again we find this guy named Satan, also known as the dragon, in the lead to overtake heaven and make it more Carnal. Let's look at Revelation and see what it describes. John saw what is believed to be an overview of the beginning of the world in a dream.

*Revelation 12:3-9 And there appeared another wonder in heaven; and behold a <u>great red dragon</u>, having seven heads and ten horns, and seven crowns upon his heads.-And his tail drew the <u>third part of the stars</u> of heaven, and <u>did cast them to the Earth</u>----And there was war in heaven: Michael and <u>his angels fought against the dragon; and the dragon fought and his angels, and prevailed not</u>; neither was their place found any more in heaven. And the great dragon was cast out, that old serpent, called the <u>Devil, **and** Satan</u>, which deceiveth the whole world. <u>He was cast out into the Earth, and his angels were cast out with him</u>.*

A <u>third of the heavenly host followed Satan</u> and the <u>dragon and they were cast out of heaven</u> because they revolted. These cast-outs became human during the Tertiary period and they were the rulers of much of the land. They were known as the Anak by the Jews, Annunaki by the Sumerians, Lords of Amenti by the Egyptians, the Araya by the people of India, and the Akamim by the Mongulala of Brazil, but they were simply giant people who had become angels and had been turned back into cursed humans.

Melchizedek- 1ˢᵗ Century BC **Jewish Essene Text-***Pray for the offspring of the angels, together with seed which flowed forth from the father of all who made the entire universe from nothing there were engendered the gods [Anak people] and angels, and the men that came out of the seed, all of the natures, those in the heavens and those upon the Earth—now the nature of females was wanting among those that are in the heavens. They were bound with men and women, but they were not the true Adam or the true Eve.*

This verse talks about a difference between angels and humans called gods {Anak} and infers that a union between normal people and the Anak people was accomplished. It specifically indicates this human was not the true Adam and Eve.

Incan History- *During the age of the giants* [This is talking about the same giants as Genesis 6:2], *a huge war broke out. The war between giants and gods ended in complete destruction.* [The world became void and without form.]

Today we have a much better appreciation about what the SOUL is. It is the essence of all humans and it does not die just like DNA chemicals continue to look like DNA after the "body of a person" is dead, the soul is now believed to be the part that holds our reality and this change from human to Angel to human is both possible and explained with science.

The time was near the end of the Cretaceous period. The reason we use this time is that its timing is right and the destruction of the Cretaceous extinctions was almost as horrible as the Permian Extinction thousands and thousands of year before so they would not match up with the Biblical story.

For the most part we can believe the world was a pretty nice place to live during the Cretaceous, but all of a sudden there was an explosion that would cause annihilation.

A meteor hit the Yucatan causing a 180 km diameter crater.

The Earth split open and magma gushed out for hundreds or thousands of years. While it was difficult for scientist to come around about this massive split in the Earth, most research papers today add intricate details and even blame the massive destructions on the magma spewing event rather than the Yucatan Meteor itself.

The magma, continuing to spew out, filling the air with iridium dust that laid a later of chalk around the entire planet over time. Researchers found that many of the dinosaurs that died during this time were found UNDER the iridium dust so their theories of the Yucatan Meteor bringing enough Iridium to encircle the Earth didn't work. The death of all the dinosaurs previous to the explosion points to devastation before the devastation. The image below shows how the entire country of India was made as millions of cubic meters of magma spewed out for a thousand years.

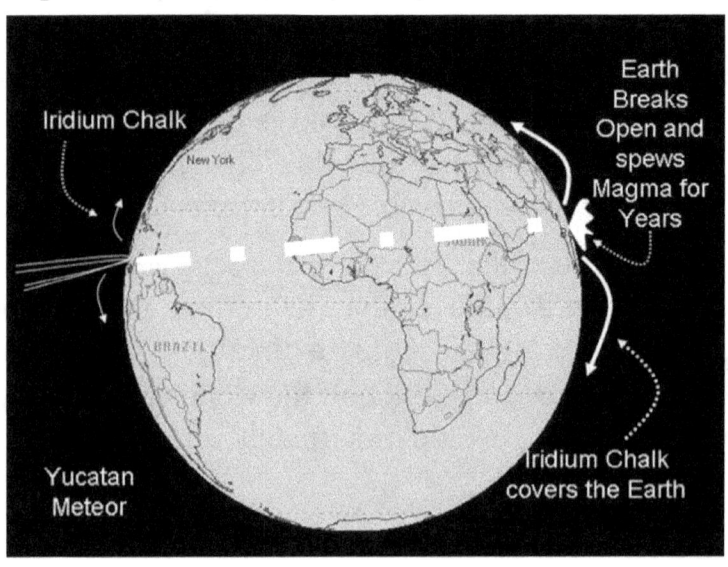

While the Earth seems to have shifted on its axis for a time, it soon settled back to its "normal" spinning. The Earth would be stable throughout the Tertiary and most of the Pleistocene, as the Anak people begin rebuilding and a new human creation appears. Unfortunately for the Anak they had been punished by losing their spirit so that they could never go back to heaven. [This is going to be bad for us later!]

Melchizedek-Pray for the offspring of the angels, together with seed which flowed forth from the father of all who made the entire universe from nothing there were engendered the gods [ANAK people] and angels, and the men that came out of the seed, all of the natures, those in the heavens and those upon the Earth—now the nature of females was wanting among those that are in the heavens. They were bound with men and women, but these were not the true Adam or the true Eve. This verse talks about a difference between angels and people called the ANAK and infers that a union between man and one of Anak was accomplished. It specifically indicates this Half-breed or Gentile humans were only partially from the lineage of Adam and Eve. They also have a little tainted blood form the Anak people, but that is a different story. Let's first look at the Angels that stayed with God during the Heaven War and see if there is any proof of them existing.

Angels

Sometimes what appears to be miracles, we are starting to catch on video for verification. Somethings this shows that a being enters this reality briefly to comfort or protect someone and then leaves just as quickly just like descriptions of angels in Judeo-Christian writings. I don't want to spend a whole bunch of time on angels and how they are still around and still helping people but I thought a few examples might be interesting. The images following are from a traffic camera.

The man on the bike runs right out in front of a semi. Just as he was about to be killed a "man" appeared and seemed to cause the man and bike to glow and finally disappear just as the truck passes. Behind the truck on the side of the rode the bike reappears along with the driver and the mysterious man. Oddly his hands are glowing. The man puts his glowing hands in his pocket and walks across the rode as the truck driver goes to the middle of the intersection. The mystery man continues to walk and vanishes while still in view of the camera as the man that was on the bike walks to the side of the rode and sits on the ground in shock. We don't have to call him a guardian angel. We can just say he was this really fast disappearing guy with glowing hands that happened to be at the right time to save an innocent.

While many angels were not what you would call angelic as the book of Jude tells us many *"left their first estate"* and I Corinthians tell us that *"women should cover their heads so that angels will not be distracted."* We mostly hear about very caring angels comforting people, especially when they are dying. The images following were taken by security cameras and seem to show shadowy people comforting horribly injured people from a wreck or mishap. Maybe the Bible is correct.

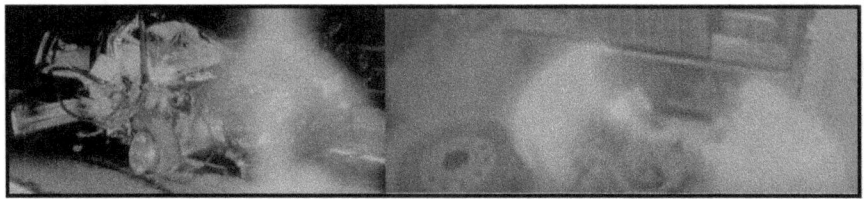

What Does Science Say?

Scientists simply tell us there must be people living in our linked universe [Heaven] by something called supersymmetry where anything that happens in our universe like having people must happen in the linked universe to hold a balance of energy and fight Entropy. It also states that not only do people live there, anything we do in our universe affects them. If we

111

gain a "soul", Heaven must lose a "spirit" to keep everything balanced and working. It's more like the workings of a machine than a mystical collection of religious questions. The more we find out about our universe and its universal laws, the more the Biblical statements thought to be either fanciful or miraculous appear to be the laws of reality.

All the ancient giant people had sort of died before and during the Heaven War. Those who had "died" became entities living in Heaven. The war separated those who would stay in Heaven and those sent back to this universe as a new people called the Anak. No long after this separation, scientists tell us, that ape-man Homo Habilis was the main primate in the jungles when Homo-Erectus man just appeared.

Sixth Age Man Confusion

Genesis tells us the second group of humans was created during the 6th Age to REPLENISH the population of people who had died as a result of the extinction described in Genesis 1:1. As I have described from other text, Genesis confirms the 3 creations of humans, indicating after he made the 6th Age man. Scientist called this human Homo-Erectus. Haplotype scientists indicate that this new individual had his first DNA mutation about 150 thousand years ago and around the world the names "Adam and Eve" were assigned to the Mutation [A,I] even though the Adam of the Genesis story was not Homo-Erectus.

Genesis 1:27-31 So God created man in his own image,---And the evening and the morning were the sixth day/Age.

This was after the heaven war described in Genesis 1:2 so it would have been during the Tertiary period. Afterward, Moses indicated God rested for a season [the 7th day/age] so we must wait until the Pleistocene Age.

Erectus Completely Different than Habilis

What we know is that there was a huge difference between the ape-man called Homo-Habilis and the human Homo Erectus. The undirected Evolutionist quasi-Scientists can't understand what happened. It was not a massive electrical storm that enhced humaoids in opposition to Entropy. Here are some of the changes noted.

113

	Habilis-ape-man	Erectus-man
Location	East Africa only	Eurasia and Africa
Build	robust - 4 feet	Taller, slender -5 feet
Face	Protruding face, prominent cheekbones	flatter face, large brow-ridges
Limbs	long arms, short legs	Slender arms and legs
Walk	Feet outward	Feet more straight
feet	Hand-like	More foot-like
Posture	Stooped	upright posture
Teeth	Large and elongated	Shovel-shaped, smaller
Brain	600 cm3	1000 cm3
Tools	Scavenging tools	Hunt & defense tools
Fire	No use	fire and fire-making
Speech	no	yes

Homo Erectus was the first true human of this line. His features were man-like including his teeth, pelvis, and legs. He was much larger and his brain had swelled to almost twice that of his very recent predecessor. Those things didn't happen by chance and they all point to a very strong, very articulate, very manlike worker. The brain expansion was almost like the brain had gone into an evolution jet going thousands of years in the future while the same general shape and appearance looked almost like a reasonable progression. Don't confuse evolution with NEW CONSTRUCTION. I know you are going to say what is scientific about that? Well the undirected evolution idea that many types of sugars just happened to be on a stump at the same time and lighning struck forcing the sugars into clumps of DNA while on another dozen stumps the exact same thing happens in exactly the same way so that a viable procreative entitity could be established sounds much more fanciful. When we add in the Soul and spirit components required for Anthropic development of our universe and the problem with Entropy alwaysforcing lower levels of being rather than allowing survival of the fittest as suggested by some and you have an entire fairy tale. After the creator God made this new being, the Anak people immediately began trying to change this less than useful servant and an explosion

114

of different type of homo-Erectus and Neanderthalis humans popped up everywhere. The following chart shows just some of the new humans.

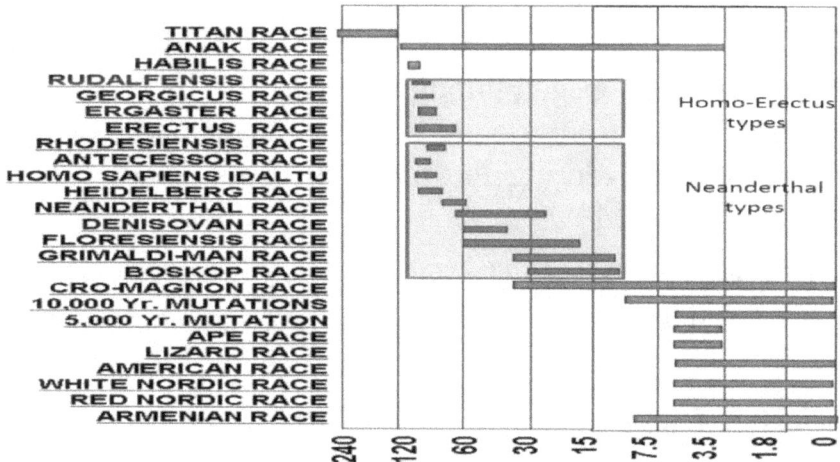

The following chart shows where the majority of the various human types were found and how these people spread around Eurasia and Africa. While the some stayed in Africa, later variants moved to other locations.

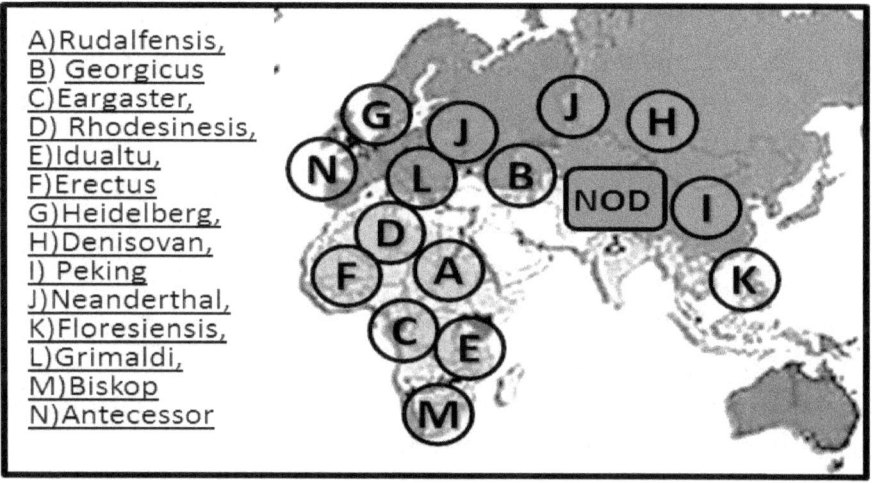

From a whole raft of Judeo-Christian, Sumerian, Babylonian, Dravidian, and similar texts we find details of the 6th day man and haw he was changed by the hand of the Anak people. As described in several books Homo-Erectus was created to be a worker. This is proven by examination as his bone structure and articulation level tells us that. The Anak people now wanted something different. They wanted to be worshipped like gods and homo-erectus groups were not worshipping the Anak people. The Anak actually thought they wanted smarter humans so they began modifying him to be smarter and stronger. Details from around helps us understand more about this man created during the 6th Age. [This is not Adam of Genesis chapter 2.]

India Creation Myth-The *first humans were covered with thick hair.* *When they mated they produced people as they are now.*

Sumerian Gilgamesh Story--*Aruru [God] pinched off some clay and created a [primitive man] Enkidu His whole body was covered in hair.* *He knew neither people nor country; with cattle he quenched his thirst, a hunter and brigand—She [Shamhat –one of the Anak] must take off her clothes and reveal her attractions. Do for the primitive man, as women [Anak] do. She pulled not away, Enkidu was aroused.* [This is the most descriptive and shows that there were female Anak that had sex with male humans. They sort of forced themselves to sleep with this lowly people.] ---*Afterward- the gazelles saw Enkidu and scattered, for Enkidu had stripped--- his body was too clean [the hair was all gone]. His legs were diminished-he could not run as before, he had become wiser. —Enkidu, you have become like a god. [Anak] -* He shall *bring up daughters of gods [hybrid man].* [The Union between the Anak and the new human produced viable offspring. We can believe this was the Neanderthal human.]

Mandeans of Iran Story-*According to their traditions, the gods first made man.* *When he was finished, he looked like a man, but moved on all fours, had the face of an ape, and made noises like a sheep. Only later did he put in a soul and teach him and make him erect.*

In Africa, the same story was told. *Hairy men became human* after coming in contact with the Anak people.

Southeast Asian history-An extremely hairy human "female" named Bota Ili was cooking food. A non-hairy fisherman named Wata Rian saw her and got her drunk. While she was asleep, he shaved her entire body. Only then did he find out she was a woman. She learned to wear clothes, they married, and they began a new race.

Emerald Tablets [Egypt]-*The master said-take them by the arts ye have learned of far across the waters until ye reach the land of the hairy barbarians, dwelling in caves of the desert. Follow there the plan.* [The plan was to inbreed with the hairy barbarians.]

Ngombe Tribe Folklore-"*A sky person [Anak human] saw a hairy man [6th day man]. She married him and removed his hair. Then a Garden was made for man to live in.*" [The offspring of this union was less animal like just like all the other descriptions.]

Mayan History- "*Men were once hairy and monkey like.*" *Something happened to them.* [I think you know what it was. It had to do with Anak and sex.]

The Anak Scientist modified all types of animals to make new and exotic creatures including modifying Homo Erectus. While the genetic scientists of that time were fantastic at gene splicing, they did not understand much about the soul and spirit of life. The animal characteristics got worse and worse and soon God was instrumental in destroying the earth. First we have to bring Adam and Eve on the scene.

Haplotype Tracking

Thanks to the scientific tracking of DNA mutations we know generally where the original Homo-Erectus people lived and where they went. Generally speaking before about 50 thousand years ago, most were located in Africa with very few mutational differences. Male tracking describes 4 mutation [called A, B, C, D for shorthand description] and female DNA tracking had 4 mutation types [called L, L_1, L_2, L_3] around 50 thousand years ago the "D" haplotype may have ventured into Europe and that would seed a new group of people called Neanderthal. The early Haplotype tracking is shown below.

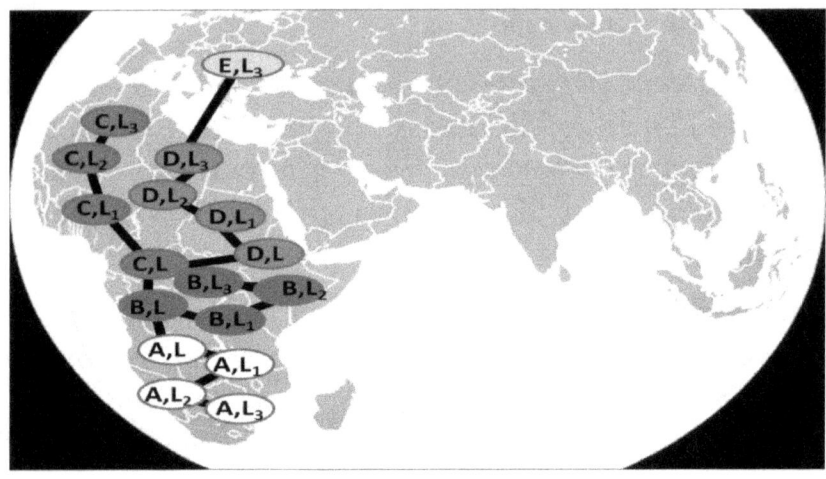

Cro-Magnon of Genesis

First let me just say Haplotype scientists tell us Cro-Magnon humans were not related to Homo Erectus or Neanderthalis before the Pleistocene Extinction 10 thousand years ago but some individuals who are related have Cro-Magnon DNA. This means there were two separate human lineages. One starting with Homo-Erectus out of magic and a second line that mysteriously sprang up in the Middle East by some unknown method. For years undirected Evolution Theories tried their best to meld some type of expansion disregarding DNA, but then we found out that if Homo-Erectus was the fabled 6th Day man and Cro-Magnon was the fabled 8th Day man/Adam, the story presented by Moses was unbelievably insightful and truthful. The other evolution ideas that continue to violate the Law of Entropy to make more and more complex creatures were fully threatened.

The earliest known remains of Cro-Magnon are about 20 to 40 thousand years old, while Homo-Erectus claims ages from the tertiary period and there is little doubt that Neanderthal came from Homo Erectus, however there are strange mutations in Neanderthal that made him more intelligent, stronger, and more Cro-Magnon like. We can assume these mutations were from the Anak people. All that being said, coming from the Middle East, Cro-Magnon and subsequent hybrids are found all over the world with a curiously small population in Southern Africa. Cro-Magnons were robustly built and powerful. The body was muscular; the forehead was fairly straight rather than sloping as found in Neanderthals and he had almost no brow-ridge. The face was short and wide and the chin was prominent. The most interesting thing about Cro Magnon is his brain capacity was about 1,600 cc; about 10% larger than either Neanderthal or modern man. [See Graph]

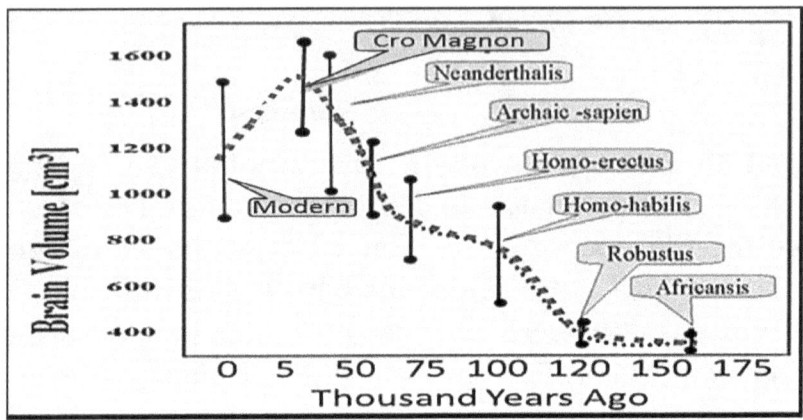

Later we will talk about what happened 5 thousand years ago to make our brain atrophy or shrink, but right now let continue. Cro-Magnon was crafty from the start and they have been found with numerous tools along with pieces of shell and animal teeth in what appear to have been pendants or necklaces. Great artists, works in ivory, and wall paintings were popular as shown below. They buried their dead intentionally showing knowledge of ritual and healed wounds show they protected family members.

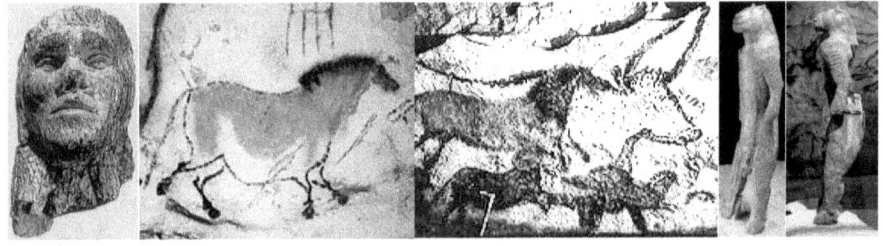

For the next detail I must tell you a little about Haplotyping. Essentially it uses mutation position and type to classify and time the various elements of an individual's heritage. There are 2 versions of DNA in a person's body [Y-Chromosome] and [chromosome associated with something called the

120

Mitochondria] that used to be a bug, but now mitochondria are in every cell of your body with their "alien" DNA. Very strangely, only 1 to 4 percent of the DNA in Europeans and Asians may be derived from Neanderthals so Neanderthal sort of dead ended with the Cro-Magnon. Cro-Magnon people are identified by Y-Chromosome DNA mutation [F] and mitochondrial DNA mutation [N] "Haplotype". This is written [F,N]. Homo-Erectus is identified as [A,L] and Neanderthal as [D,C1] the last two type of people descended out-of-Africa. The simple Haplotype tree below shows lineages.

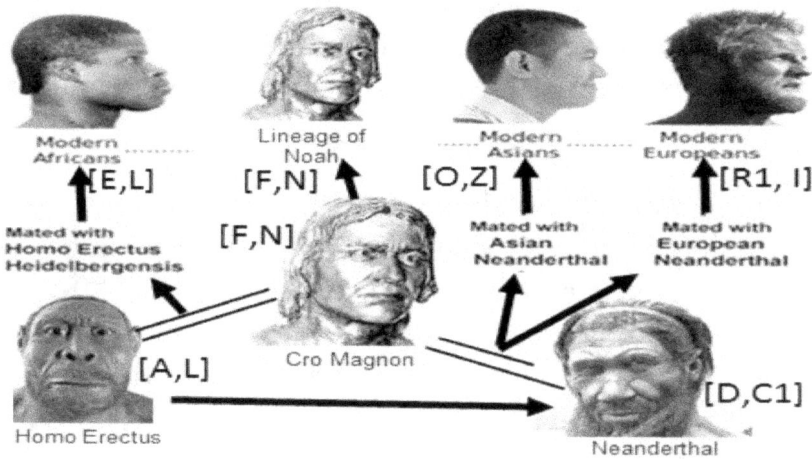

In a resent Haplotype study of almost 8 thousand individuals showed European lines had no A or B "homo-erectus" DNA mutations showing there was no early African link and almost no Neanderthal.

Bible Description

The Bible indicates a third human was created during the 8th Age. Scientists tell us Cro-Magnon humans simply appear one day at the beginning of the Pleistocene Age well after the Homo-Erectus humans of the Tertiary period. Neither science nor the Bible can separately answer us adequately, but together, the answer becomes clear. DNA tells us Cro-Magnon did not "Come out of Africa" so from where did he come? In fact, the researchers made note of their repeated

121

absence stating "not one non-African participant out of more than 400 individuals in the Project tested positive to any of thirteen 'African' sub-clades of Haplotype A". The only remaining uncertainty relates to the identity of this "more ancient common ancestor". All that can be stated with confidence is that humanity's ancestor did not reside in Africa. Unfounded accusations of racism have become common as the prevailing Afrocentric hypothesis is constantly being challenged by the growing mountain of conflicting scientific evidence, especially in the evolving field of genetics. All Cro-Magnons are Modern, but all Moderns are not Cro-Magnon Average stature of the Neanderthal Man was about five feet four inches, whereas Cro-Magnon Man averaged over 6 foot.

Retiming Adam

It is true that many Jewish and Christian groups try to explain that the Bible indicated Adam lived only 930 year and he must have been created on "October 23, 4004 BC at 9:00 AM". I have no idea if that was daylight savings time or not, but there are questions about that date. As I mentioned before, the Yawm/day word in Genesis simply means age, but the question is. "Can the time Adam was created be placed on the time that Cro-Magnon was created or were there 2 different creations? If we look at 2 different Jewish texts; Adam and Eve I and Adam and Eve II we find that Adam was possibly promised to live for another 5500 years after the convenient was made rather than the 930 years interpreted in the book of Genesis.

Adam and Eve I 3:2--Yea the word that will again save thee [Adam] when the five days and a half are fulfilled – the God in his mercy explained to him that these were 5000 and 500 years and how one would come to save him and his seed.

Adam and Eve I 21:9--And [God] said to Adam, "O Adam all the misery which thou hast wrought upon thyself, will not avail

122

against my rule, neither will it alter <u>the covenant of the 5500 years</u>.

Adam and Eve I 38:2- *O Adam, as to the fruit of the tree of life, for which thou askest, I will not give it thee now, but when <u>5500 years are fulfilled</u>. Then will I give thee of the fruit of the tree of life, and thou shalt eat, and live forever, thou, and Eve, and thy righteous seed.*

Adam and Eve II 19:1—*Then God revealed to him [Adam] again the promise he had made. He explained <u>to him the 5500 years</u> and revealed to him the mystery of his coming to earth.*

Generational Dating

Knowing Jewish timing is generational rather than yearly and that early Years were actually called "Jubilees" or 7 year time periods, we simply use the relation rather than the specific date. If we simply convert years to Jubilee Years, we get the following chart. For confirmation Adam was told the 5500 year convenient time after Shem had been born so he was about 1000 years old by the chart below and the 6500 year death matches the details of the references provided. While this indicates Adam was born about 21 thousand years ago. It is pretty reasonable and it is triggered on the end of the Pleistocene age as the time of the worldwide flood and shift in the Earth which corresponds to other details.

Adamic Patriarch	Genesis son birth year	Son born year in Jubilees	New Birth [Yrs Ago]	Genesis death year	Died Age in Jubilees	New died [Yrs Ago]
Adam	130	910	20892	930	6510	15292
Seth	105	735	19982	912	6384	14333
Enos	90	630	19247	905	6335	13542
Cainan	70	490	18617	910	6370	12737
Mahalaleel	65	455	18127	895	6265	12317
Jerad	162	1134	17672	962	6734	12072
Enoch	65	455	16538	365	2555	14438
Methuselah	187	1309	16083	969	6783	10000
Lemech	182	1274	14774	777	5439	10000
Noah	500	3500	13500	910	6370	7130
Flood			10000			

123

This timing is significant in that many other pieces of physical evidence and written testimony indicate that Adam was created between 20 and 40 thousand years ago and this verifies that timing. If that was all we had I would not agree with it, but then we have the Biblical book of Ezra and many other similar dates.

4 Ezra 14: 10-12-For the world hath lost his youth, and the times begin to wax old. For the world is divided into twelve parts, and the ten parts of it are gone already, and half of a tenth part: And there remaineth that which is after the half of the tenth part. These writings come from about 2.5 thousand years ago. If we assume there is about a thousand years to go before the earth is destroyed and that makes up only 1/8 of the total time the earth has been around, the beginning of man would have been 28 thousand years before the time of Ezra or 30 thousand years ago.

Babylonian History-*King Alalamar* [one of the first Babylonian rulers] *ruled for 36,000 years before the flood.* When added to the 9 thousand years since the flood, this king, ruled Babylon about 45 thousand years ago.

Manteo's "History of Egypt"- Gods ruled for 13,777 years followed by 15,150 years of rule by demigods and spirits of the dead before the flood. When added to the 9 thousand years since the flood, the first man or Gentile ruled Egypt 39 thousand years ago,

Turin Papyrus- Gods ruled Egypt for 13,420 years followed by 23,200 years of rule by demigods. When added to the 5 thousand years since the papyrus was written, the first man or ruled about 41 thousand years ago.

Emerald Tablets- According to this Egyptian text, *Thoth was sent to Egypt from Atlantis before it sank. He was sent 37 thousand years ago; he became the ruler; and at some point, he created the first 2 pyramids.* Assuming the 30 thousand years started near the time of Adam and Eve, and the reference is about 5 thousand years old, then Adam and Eve would have been here about 43 thousand years ago.

Byzantine History-Syncellus wrote in the 9th century AD, that *the chroniclers of the pharaohs had recorded events for 36,525 years.*

Please remember that this whole dating thing is arbitrary as Anthropic reality makes its own timing to support the NOW. I simply put together a way to not have the anomalies we have in scientific understanding of the past.

Haplotype Verification

Besides the Bible, there are many texts that verify what we are told by Haplotype DNA studies. Homo-Erectus and Cro-Magnon were 2 separate creations.

Mandean/Essene- "Iranian Folklore"-First God made man [Homo-Erectus]-*Only later [during the Bronze Age] did he [God] put in a soul and teach him and make him erect. The soul entered Adam and he stood erect and talked and God taught him reading and writing and all knowledge.*

Yezidi/ Kurds-"*Mishaf Resh*"-*God made man [The Homo-Erectus had been made in the Silver Age]* then *He* [God] *created Adam and put in him a soul. On Saturday, He created Jibrail [a new Archangel to replace Gadrael/Satan] as a substitute for Tawus Malak. He made him head over them all.*

Gnostic-"*Hypostasis of the Archons*"- -*The rulers laid plans and said," Come let us [the Anak] make man that will be soil from the Earth."[Homo-Erectus man] They molded their creature wholly from the Earth. Now they made a body. They modeled that man after their own bodies. They did not understand the force of God. God breathed life into the man's face and he became a living soul. The man was called Adam.*

125

God made Adam sleep and opened his side like a woman. In this version Eve was "attached" to Adam

Gnostic- *"Creation Text"-* *[The second Adam was Homo-Erectus].The third Adam is a creature of the Earth, that is the man of the law. He appeared on the eighth day/Age and became numerous and produced every kind of scientific information of the soul endowed Adam."* [We already read about the first 2 Adam's created by Incarnate God. This one would be Cro-Magnon. The 8th day is code for beginning of the Pleistocene.]

Haplotype DNA Mutation science indicated Cro-Magnon just appeared one day and other details indicate this was the famed Adam that Moses wrote about. The mystery of how Neanderthal got so advanced from Homo-Erectus is a mystery to blind scientists, but it makes total sense once the Bible is used as a source of scientific information as there was a large number of scientists during the Pleistocene Age that the Bible indicated worked on Genetics.

Ancient Genetics

Around the world and in our Bible we read about unclean or "abominable animals". The reason most of the animals of this time were considered abominations was a massive thrust for geneticists to modify animals. Besides our Bible, we find out from other Jewish texts the same thing as sceintific research in Genetics and Engineering bring us to a dangerous level. I know this sounds like some bizzare fiction, but bear with me for a minute as we review a tiny portion of the Judeo-Christian texts concerning this seemingly erroneous fact pushed in the Bible.

Book of Giants- - they knew the secrets of heaven and sin was great in the Earth. They made mistakes] and they killed many [animals and people.]- [They selected two hundred] donkeys, two hundred asses, two hundred [rams of the] flock, two hundred goats, two hundred [other beasts of the] field. From every animal, and from every [type of human was taken its seed] for mixed sex. [After a time] they defiled the animals and people and begot giants, monsters, and dragons.

Book of Secrets- Those who would penetrate the origins of knowledge, along with those who hold fast to the wonderful mysteries of life. -With this I beseech your attention. All of the secrets of manipulating life were known but they [the ancient humans] did not know the secret of the way things are nor did they understand the things of old. Belial who modified creation, a thing that ought never to be done again, except by the command of his Maker. You have not become wise in understanding my secrets of life and the earth; for you have not properly understood the origin of Wisdom.

Jasher 4:16-18- and the sons of men in those days took from the cattle of the Earth, the beasts of the field and the fowls of the air, and taught the mixture of animals of one species with the other, in order therewith to provoke the Lord; and God saw the whole Earth and it was corrupt, for all flesh had corrupted its ways upon Earth, all men and all animals. --- And after this they sinned against the beasts and birds, and all that moves and walks on the earth:

Enoch 2:4-5 And the sons of men went and they served other gods, -- and the sons of men forsook the Lord all the days of Enosh and his children; and the anger of the Lord was kindled on account of their works and abominations which they did in the earth—

Enoch 4:8-And lawlessness increased on the Earth [War] and all flesh corrupted its way, alike men and cattle and beasts and birds and everything that walks the Earth all corrupted their ways and their orders. The only way that animals corrupted their way was that they were genetically manipulated and just weren't the same animals.--

Enoch 7:5- And they began to sin with birds and with animals and with reptiles, and with fish. [This did not mean that the rulers had sex with fish. This is talking about manipulation of species]--

Jubilees 5:3--and all flesh corrupted its way, alike men and cattle and beasts and birds and everything that walks the Earth all corrupted their ways and their orders. [The only way that animals corrupted their way was that they were genetically manipulated and just weren't the same animals.]

Jubilees7:24- Afterwards they sinned against beasts and birds and everything that moves or walks upon the Earth. [There are

128

two ways to sin against beasts- sex and genetic manipulation. God didn't like either.]

Enoch II 59:5-6- But whosoever kills a beast without wounds, kills his own soul and defiles his flesh. And he who does any beast any injury whatsoever, in secret, it is evil practice, and he defiles his own soul. [The killing and injury done in secret was not killing animals for food, it was the genetic manipulation and corruption by integrating man's; genetic material.]

Generations of Adam 6:1-10- AMONG our little ones - Timnor's sister Ammah was also blessed with understanding, for she investigated the nature of life, unlocking the mysteries of life itself. -- Ammah was not one whit behind her husband in creating wickedness, for she manipulated the very fountain of life, until she had created new forms of beings dedicated to evil and the destruction of mankind.

Generations of Adam 8:4-20 Timnor and Ammah practiced every abomination. Tranter learned the ways of his mother Ammah and he did manipulate the very nature of man and beast to create new forms which God had not ordained.

The Zoroastrian "ZAND-AKASIH" - *Satan miscreated creatures and they became useless. God saw the defiled and bad creatures, they did not delight Him* [They became abominable]. *Satan's downfall was the unrighteous creation of the creatures and ignorance.*

Jurassic Park

Today we have fantasy movies about Jurassic park/ remade animals and we are finding out that during the Pleistocene Age, many animals were not only made by gene splicing, but also ancient animals were re-made to be abominations to God. We are finding dozens of unfossilized dinosaur remains showing they were re-made less than 40 thousand years ago even when it has been determined that the Duckbill dinosaurs, Tyrannosaurs Rex, and other unfossilized dinosaurs did, in fact, become extinct by the end of the Cretaceous. Someone must have found viable DNA almost like the Jurassic Park movies. Speaking of wrong animals in the wrong era, check out the next section. Now the world is being flooded with soft tissue, dinosaur blood, dinosaur skin, and all sorts of unfossilized dino-stuff. Some of the blood and tissue samples are next.

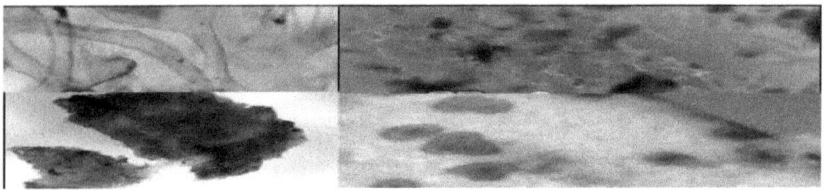

Besides blood, researchers are finding all sorts of pliable, soft tissues. Some colleges have even tried to recreate the same type of experiment identified in Jurassic Park movies. Below are more samples.

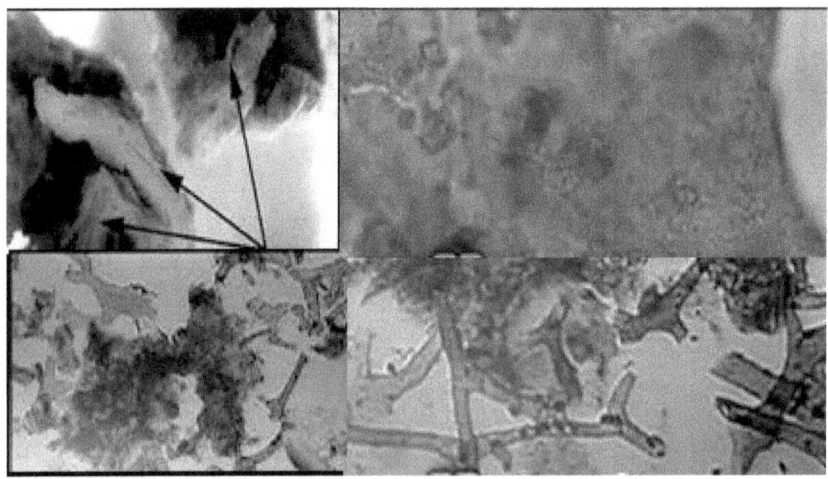

Apes Before Dinosaurs?

Why do the ape skeletal remains seem to have such a horrible look on its face? [See next left] The reason is obvious to a casual observer, but hard to accept by the ardent evolutionist, because, the fear experienced by this mammal was the fear of being eaten by an allosaurus well before it could have become evolved to the high form of the ape. The find was in upper New Mexico, but typically, you don't see these things in text books. One of the photographs of this impossible scene is shown below and no, it was not produced with trick photography.

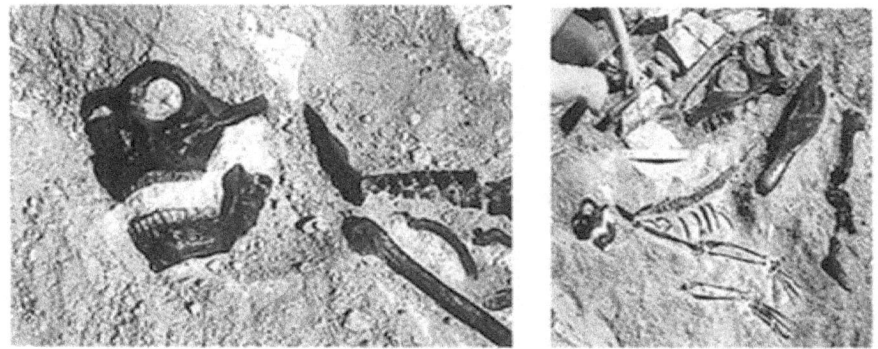

Entropy of Evolution

Just think about this. If scientists would just read some of the ancient texts they would not have to hide the anomalies any longer and there would be a logical reason why people remembered seeing dinosaurs and painted then, carved them, and etched them into wall.

| Palestine | Greece | Utah | Colorado |
| Guatemala | Peru | Mexico | Cambodia |

The images above from Palestine, Greece, Utah, and Colorado, Guatemala, Peru, Mexico, and Cambodia show

131

manmade dinosaurs were one the earth even past the Pleistocene Extinction. An additional group from Peru, England, Spain, Babylon, Israel, and Egypt show that dinosaurs were not rare for a while.

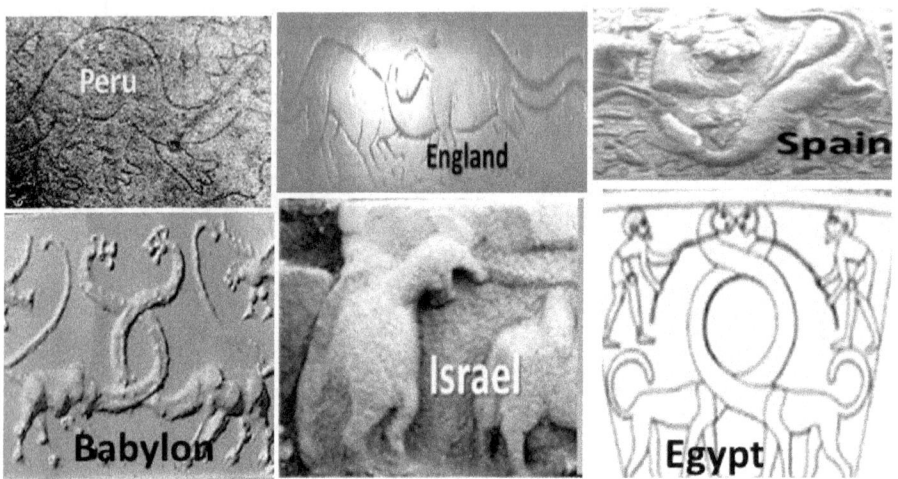

Then there is the book of Daniel in the Bible. This is the last chapter in the King James Version of the Bible. In it King Cyrus thinks his captured dinosaur is a god that can't be killed so Daniel kills the beast. Later versions have taken this part out of Daniel because they simply didn't know about all the other evidence.

Daniel [Bel and the Dragon] 1:23-28- *And in that same place there was a great dragon, which they of Babylon worshipped. And the king said unto Daniel, Wilt thou also say that this is of brass? lo, he liveth, he eateth and drinketh; thou canst not say that he is no living god: therefore worship him.-Then said Daniel unto the king, I will worship the Lord my God: for he is the living God. But give me leave, O king, and I shall slay this dragon without sword or staff. The king said, I give thee leave. Then Daniel took pitch, and fat, and hair, and did seethe them together, and made lumps thereof: this he put in the dragon's mouth, and so the dragon burst in sunder: and Daniel said, Lo, these are the gods ye worship. When they of Babylon heard that, they took great indignation, and conspired against*

132

the king, saying, the king is become a Jew, and he hath destroyed Bel, he hath slain the dragon, and put the priests to death.

There is no question that evolution occurs but in order for modifications to not push a species further and further into disorder [Law of Entropy], some outsider MUST modify the chain of events to assure advancement of a species rather than the slow destruction of entropy. In this case the overseer might be long forgotten genetic scientists or God himself, but today we know evolution is directed. If someone would have just read the Bible before trying to establish a theory, they would certainly have been better off and not have to lie so very often or hide evidence that goes against some idea based on trying to disprove the Bible. Let me give you a for instance. The War described in the Judeo-Christian texts that occurred before end of the Pleistocene extinction and worldwide flood has been almost removed from view. The radioactivity levels, the massive destruction, the huge amount of human mutation are just considered anomaly by some. Let's look at the Pleistocene war.

Pleistocene War

Everything was going reasonably well for scientists until they started trying to determine what happened towards the last part of the Pleistocene. They discovered signs of war and nuclear fallout during what is laughingly called the Young Dryas [10 to 12 thousand years ago]. No-one thought to look at the Bible to help them develop a scientific explanation and the signs of civilization around the world were hidden rather than used to understand our world better. Let's just read a few verses and try to make sense of the details uncovered by scientists.

Genesis 6:5-7-And God saw that the wickedness of man was great in the Earth, and that every imagination of the thoughts of his heart was only evil continually. - And the LORD said, I will destroy man whom I have created from the face of the Earth; both man, and beast, and the creeping thing, and the fowls of the air. [Moses soft sold the whole war episode but let's continue.]

Jasher 2:5-6- "-and the sons of men forsook the Lord all the days of Enosh [Adam's grandson] and his children; and the anger of the Lord was kindled on account of their works and abominations which they did in the Earth. And the Lord caused the waters of the river Gihon to overwhelm them, and he destroyed and consumed them, and he destroyed the third part of the Earth, and notwithstanding this, the sons of men did not turn from their evil ways--" What this is saying is that 1/3 of the world population died BEFORE the famed worldwide flood. They died in war, but the details are still sketchy.

III Baruch 16:2-*[God said] Let not the sons of men alone, but since they [Anak] angered me in their works, go and make them envious and provoked against a people that is no people, -and punish them severely with the sword and with death.*

Jubilees 5-7-*And against their [Anak] sons went forth a command that they should be smitten with the sword----And he sent his sword into their midst that each should slay his neighbor, and they all began to slay each other till they fell by the sword and were destroyed from the Earth.--And lawlessness increased on the Earth and all flesh corrupted its way, ---and the Giants [half-breeds] slew the Nephilim [Anak], and the Nephilim slew the Eljo [half breeds], and the Eljo mankind, and one man another.*

Enoch 10:13-14-*Destroy the children of fornication, the offspring of the angels, from among men; bring them forth, and excite them one against another. Let them perish by mutual slaughter.*

Generation of Adam 11:3- *[During the war] Leboa, Daughter of Tamar, devised a "Sword of Light" which penetrated the wall of defense around the city of Haner and began to drain the power from the wall.*

Incan History- *During the age of the giants, a huge war broke out. The war between giants [Eljo] and gods [Anak] ended in complete destruction.* According to the Inca, the age of giants was during this time.

Scientific Evidence

Forensic Science-Some scientists started looking for holes and found bullet holes in skulls of Cro-Magnon and earlier people, pointing to war during the Pleistocene. Here are just a few.

135

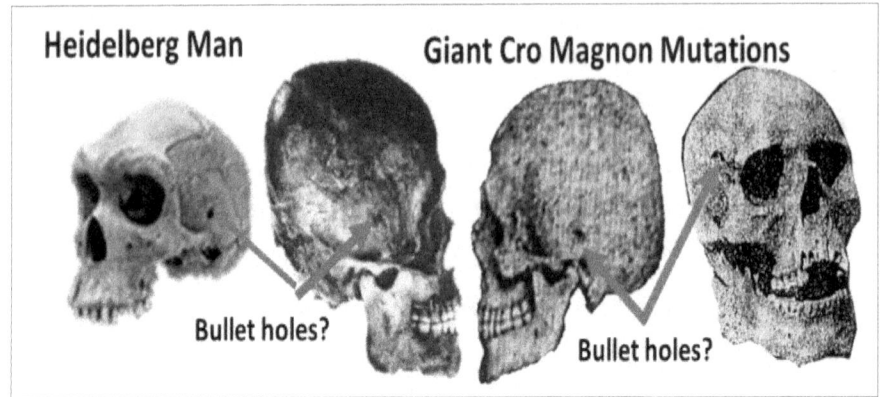

Heidelberg Man   Giant Cro Magnon Mutations

Bullet holes?   Bullet holes?

In Russia and extinct Auroch was found with a similar high speed projectile hole in its forehead which is believed to have been put there during the Pleistocene. A similar hole was found in the same type animal in Zambia and other signs of hostilities. Most had a tiny hole going in and much larger exit destruction. Please notice something that might be bad. Many of the bullet holes are from point blank range given the entrance hole positions as if done by a contract killer. I wonder if it was some form of mafia.

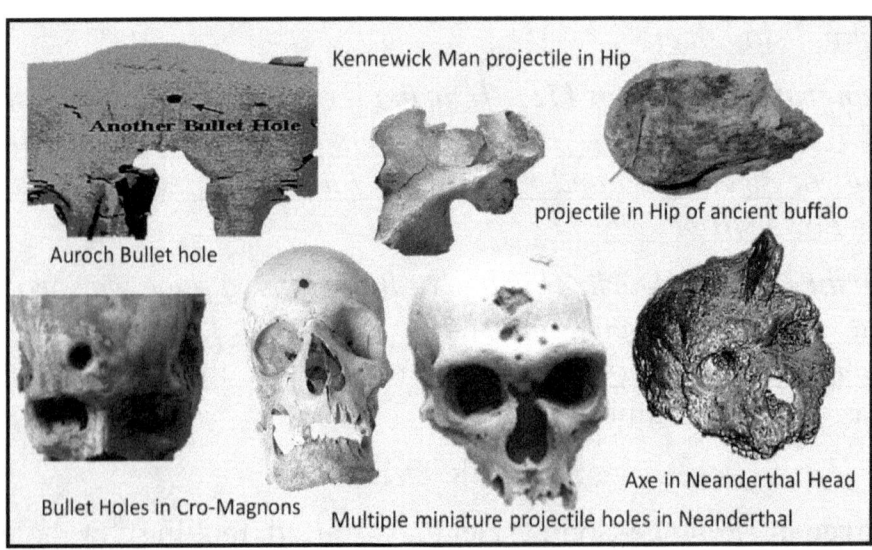

Another Bullet Hole

Kennewick Man projectile in Hip

projectile in Hip of ancient buffalo

Auroch Bullet hole

Axe in Neanderthal Head

Bullet Holes in Cro-Magnons   Multiple miniature projectile holes in Neanderthal

Nuclear Science- Large numbers of these unfossilized Dinosaur parts [especially T-Rex] are so radioactive that they must coat them with lead paint before displaying them. We

136

know that many animals and plants became extinct at this time. Then additional indicators tell us more. *Uranium concentrations in coral jump by almost 300%*. Also we find marked increases in nanodiamonds, magnetic spherules [tiny balls], and carbon spherules at the end of the War with a major increase in charcoal around the middle showing fire and general war conditions. The nanodiamonds indicate substantial heat like nuclear explosions. The darker area below represents the Dryas which occurred between 10 and 12 thousand years ago, around the same time as described in Jewish histories as the time of the massive war.

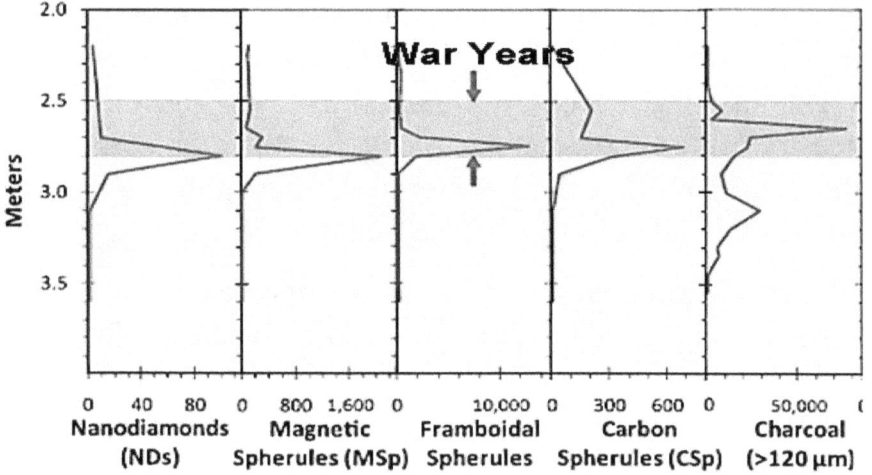

Let me tell you something about massive amounts of radioactivity. It mutates DNA.

People Mutated

I know all this war before the end of the Pleistocene and the nuclear destruction doesn't souls right, but that is what the evidence shows and it agrees with Judeo-Christian details. Another thing that is shown is that people during this time mutated as if they were somehow affected by nuclear radiation----[possibly by war]. The study of DNA is a fairly new science, but it is rapidly getting more and more detailed as Haplotyping of DNA mutation is used to find out the ancestry of just about anyone today. A study dating the age of more than 1 million single-letter mutations in the human DNA code revealed that most of these mutations are of recent origin.

- According to one study, over **86 percent** of the harmful single nucleotide mutations arose between 5 and 12 thousand years ago. Oddly, since then there have been very few major mutations.

- Overall, the researchers now believe that about 81 percent of the single-nucleotide variants in the European sampled and **58 percent** in the African DNA sampled arose in the past 5 or 6 thousand years.

- In the African samples a large number of the single nucleotide mutations appeared before Cro-Magnon [20 to 40 thousand years ago], so those would have had more Homo-Erectus ancestry, but among Europeans and the rest of the world, many mutations occurred between 10 and 12 thousand years ago.

By looking at sequential mutations recorded in "y"-sex chromosomes, the following list of races around the world whose beginnings have been fixed and decoded. Please notice most changes occurred 10 to 12 thousand years ago, during the time of the great Pleistocene Wars, or 5 to 6 thousand years ago which coincides with another war described in Judeo-Christian texts. Called the Bharata War, we will look at it a little later.

Mutation indicator	Time [x1000 yr]	Races of Men caused by mutation
Y	100	First Y-DNA
B	50	Sapien Sapien
F	40	Cro Magnon
C	20	Negroid
E	12	Nubian
G	12	Armenian
I	12	Greek
K	10	Asian
J	10	Canaanite
R	10	Scythian
H	10	Afghan
P	6	Proto Ameridian
R2	6	Aryan
N	6	Russian
O	6	Oriental
L	5	Dravidian/Indian
R1a	5	Slavic
R1b	5	Gaelic
N	5	Scandinavian

The beginnings of the Nubians, Armenians, Greek, Asian, Canaanites, Scythians, and Afghans all are traced during this horrible War. We even have references found within the

139

Scrolls of the Essene. The fragmented book is generally called "Mixture of Adam".

"Mixture of Adam"

These fragments were found in cave 4 in 2 different sections. [4Q186 and 4Q561] The book tells about the many mutations of people living during the Pleistocene. While it is not clear if these differences were from DNA mutation or inbreeding with the Anak, but there seems to be similarities between this data and the mutation characteristics found from Haplotype mutation testing.

11% Adamic Description-*4Q186 Fragment 1-*His head and *cheeks are fat; eyes are terrifying; teeth are different lengths; hands and fingers are thick; thighs are thick and very hairy; toes are thick and short. His spirit has eight parts in the house of darkness and one in the house of light.*

[25%] Adamic Description-*4Q561 Fragment* - His body hair *is ample; voice is stern and does not strain; hair of his beard is plentiful; neither fat nor thin; short in stature; nails are strong...* [The percentage detail of this race and several of the others was not recoverable so it is listed as probable only by characteristic similarity.]

[40%] Adamic Description-*4Q561 Fragment*- His beard is *reddish; eyes are clear and circular; hair of his head ...*

67% Adamic Description-*4Q561 Fragment*- His Head is *wide; chin is thin; body is tall; body hair is full; build is thin but well built; hands and feet are medium length and thin; his eyes are fixed. His spirit has three parts in the house of darkness and six in the house of light. [67% Adamic]*

[75%] Adamic Description-*4Q561 Fragment* - His hair is *mixed and sparse; eyes are of a medium shade; nose is long and attractive; teeth are straight; beard is relatively thin; limbs in fit condition and medium built; elbows are strong and* *husky; thighs are of medium bulk; feet are of medium length; shoulders are medium width…*

89% Adamic Description-*4Q186 Fragment* -*His eyes are neither dark nor light; beard is light and curly; voice is soft and gentle; teeth are fine and well aligned; size is medium and well built; fingers are thin and long; thighs are hairless; soles of his feet and toes are even and well aligned. His spirit has eight parts in the house of light and one in the house of darkness.*

To make it easier to identify the differences, I have generated a chart of the characteristics from the fragments and identified how they relate to un-mutated Cro-Magnon or Adamic people. The percentages indicated in brackets were added to show presumed numbers because that was part of the missing information. They were determined by placement in the manuscript and details. The highlighted areas were either not recoverable or simply not mentioned. I have placed possible description generalizations in those positions.

% Adamic	11%	25%	40%	67%	75%	89%
Build	Thick	Med.	Med.	Thin strong	Med. fit	Med. strong
Size	Short	Short	Tall	Tall	Med.	Med.
Feet/ hands	Thick	Strong Nails	Med.	Thin	Med.	Even/ Thin
Hand length	Short	Short	Med.	Med.	Long	Long
Beard	Hairy	Hairy	Red	Full	Thin	Curly/ Lt.
Body hair	Full	Ample	Red	Full	Thin	None
Head	Fat	Round	Round	Triangle	Oval	Oval
Teeth	Varied	Spaced	Spaced	Even	Even	Fine
Voice	Stern	Stern	Med.	Med.	Soft	Soft
Eyes	Scary	Round	Clear	Fixed	Med.	Med.
Nose	Fat	Fat	Med.	Med.	Long	Med.

While this type of characterization is skeptical at best, at least one can recognize that these early people recognized that there were mixed breeds of humans that possessed more of less of the Adamic blood. Noah's family were all full blooded Cro-Magnon [Haplotype "F"], but these other people were substantially different. If I were to guess Haplotypes of these people I would suggest the following:

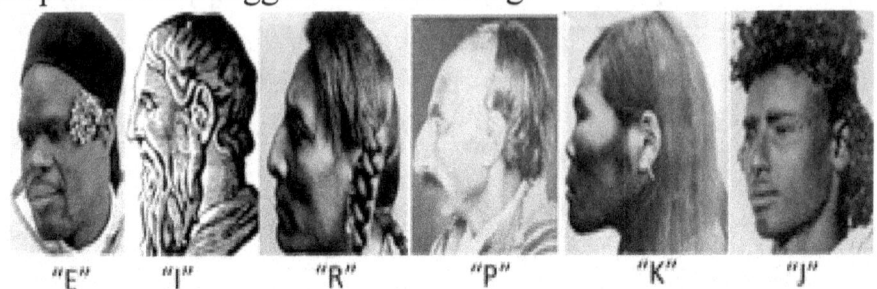

"E"   "I"   "R"   "P"   "K"   "J"

11%-proto-Nubian [Haplotype E]   67%-Proto Armenian [Haplotype G]

25%-proto-Greek [Haplotype I]   75%-Proto-Asian [Haplotype K]

40%-Proto-Scythian [Haplotype R]   89%-Proto-Canaanite [Haplotype J]

First people who lived during the Pleistocene were very civilized and professional similar to our geneticists, physicists, and engineers developed a high standard of living. The Bible and many ancient books go into great pains telling us about how most of the animals created in the Pleistocene were abominations to God as the DNA was manipulated. One group of the animals that were remade were some of the dinosaurs, another was a modification of human DNA producing some of the higher levels of apes. I could discuss a large number of texts and other evidence or you can just consider these as anomalies. The other things I mentioned were mutation and radioactive remains of some of the remanufactured dinosaurs. The ancient texts talk about massive wars and the radioactive remains suggests the use of some of the processed uranium of the prehistoric processing plants found in Gabon, Africa. While these both sound fanciful, the alternative is much more fanciful and everything is simply labeled anomaly.

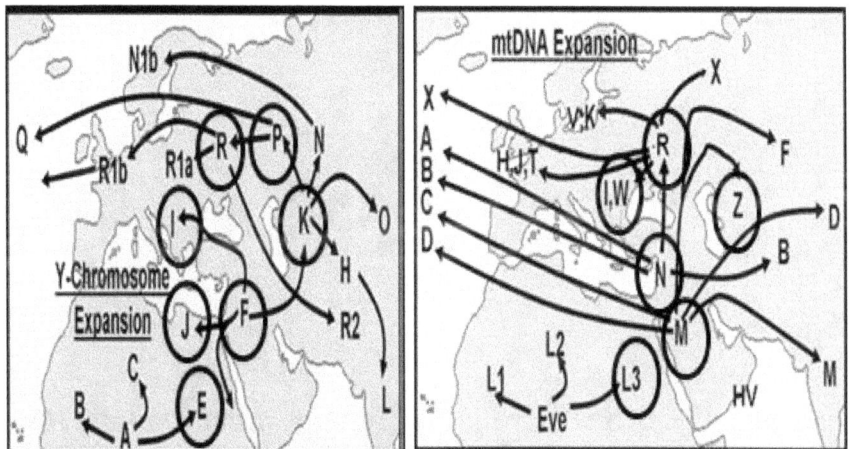

The Haplogroup map of DNA mutation tracking shows where these first mutated humans settled for a time. The map on the left is the male Y-Chromosome mapping typically used as event separation is easier to discern, but the female Mitochondrial DNA map is shown to the right to show consistency. One of the problems with scientists accepting artifacts and DNA mutations that clearly show massive wars described in Judeo-Christian documents is that they are all over the world and those not reading Judeo-Christian literature have a hard time describing how the war could be worldwide. That's where Airplanes come in. Luckily, there are all types of physical and written evidence of flying transport and commerce.

Biblical Flying

The Bible, Dead Sea Scrolls, and many other ancient Jewish and similar texts tell us over and over again that people flew and that even trips to the heavens was possible, but is there any proof While we are on the subject of flying, let's talk about the ancient Jewish people and their testimony as given in the Bible and Dead Sea scrolls. Most have read Ezekiel chapter 1 and chapter 10 about his vision of the flying wheel inside a wheel with portals all around and flames coming from beneath as it rises off the Earth, but that is only a tiny segment of the descriptions provided to let us know about this ancient mode of travel. Many of the sightings and descriptions were from before the flood. Flying ships survived the flood, however, and were continuously seen in the past. Biblical texts are full of details about this mode of travel. Before we look at some examples let's identify the terrifying flying craft or crafts as understood by the ancient Jewish people. They were commonly known as thrones, fiery chariots, whirlwinds, hashmals, and merkaba.

Fiery Chariots-The flying fiery chariots talked about in Middle Eastern histories were all similar in that they were fiery, fast, and fearsome. The early Jewish people called these chariots, Merkaba, but while a name was given to them, <u>the merkaba were so sacred that the Jews were forbidden to speak of them</u>. The early Jewish people might not have spoken about them, but they did draw them. Many drawings of Merkaba

144

have been found in various parts of Babylon and Chaldea. The writers of the day had much to say about the flying ships even though some forbade it. Notice that, in the book of Enoch, it clearly indicates that the ships went into space and that a huge ship was in orbit around the Earth. Weird isn't it? We're trying to do that now. The ship Enoch saw orbiting the Earth may have been the one or ones that eventually saved some of the ancient humans during the flood, according to the Sumerian and ancient Jewish texts.

Jubilees 33: 2-And she [Eve] gazed steadfastly into heaven and beheld a **chariot of light** *borne by 4 bright eagles- and ancient people going before the chariot* Eve watched Adam being taken into heaven on a chariot of light.

Jasher 3:36- Enoch ascended into heaven in a whirlwind, with horses and chariots of fire.

Enoch 74:15-I saw likewise the chariots of heaven, running in the world above to the gates in which the stars turn, which never set. One of these is greater than all which goes around the world. Like our current space station only larger.

Egypt and Jewish writing- The Merkaba, translated "chariot", was depicted on buildings as a set of wings which could carry people anywhere. Jewish priests forbade the mention of the Merkaba. *Elijah's fiery Merkaba took him to the heavens where he was allowed to see ancient people ascend and descend to celestial halls and to the throne of glory.*

*Midrash to Proverbs-*In this Kabalistic text *"Ishmael says, "Why have you not also studied the Merkaba, to perceive my [God's] splendor? For none of the pleasure I have in my creation is equal to that which is given me when scholars look beyond the Torah and see and behold and meditate on: My throne, and the hashmal seen by Ezekiel, and the fiery streams under my throne, and the bridges that cross it, and the Ofanim [a class of angel], and the Gilgalin [another class of angel].*

145

And is this not My greatness, and My glory and My beauty: that My children know My splendor by seeing all this?"

Letter of Aristeas- *"If it suppose that we are actually sailing on the sea in boats or flying through the air."*

Generations of Adam 6:4-5-. *Timnor [Grandson of Adam] built great engines with which to deface his Mother Earth, defiling her for his unholy purposes [During the Pleistocene]. With other machines of his contriving, his people flew through the air like birds and explored the depths of the lakes and rivers. He created also great instruments of destruction with which his people attacked the people of Cain, driving them into the mountains and dense forests, that they might take their land for an inheritance.* Men flew in the air and went under the sea during the Pleistocene War.

Egyptian Hieroglyphics- *"Osiris and Isis [Anak people] descended from the sky in a sun-ship, bringing wheat, and the arts of civilization to the world."*

Japanese Tradition- *"The spirits [ancient people] of the Shahman legends came to earth. Some of whom abandoned their ships at a respectable height and came down from them quietly on to Earth as if moving down an invisible ladder."*

India Mahabharata- *"The Gods came in their respective flying vehicles to witness the battle between Kripacarya and Arjuna. Even Indra, the Lord of Heaven, came with a special type of flying vehicle which could transport 33 divine beings."-- "He [one of the gods] entered into the favorite divine palace of Indra and saw thousands of flying vehicles invented by the Gods lying at rest"*

Sumerian Enlil Story- *"She valiantly ascends towards heaven. Over all the peopled lands she flies in her MU. To the heights of heaven, she joyfully wings. Over the rested places she flies in her MU."*

146

Sumerian Rape of Ishtar story- [Long distance flight in one day] *"One day my queen, <u>after crossing the heavens</u>, crossing Earth, after crossing Elam and Shubur, after crossing... the hierodule* [flying ship] *approached weary, and fell asleep."*

The following collage shows various flying ships from Sumeria, Babylonia, Chaldea, French cave drawings, Australia, Egypt, Turkey, and India. They have one thing in common they all were flying machines showing flying was a common travel method before the unfortunate events of the Bharata War. Notice the one next to last. It even has 3 separate rocket exhausts just like our space shuttle.

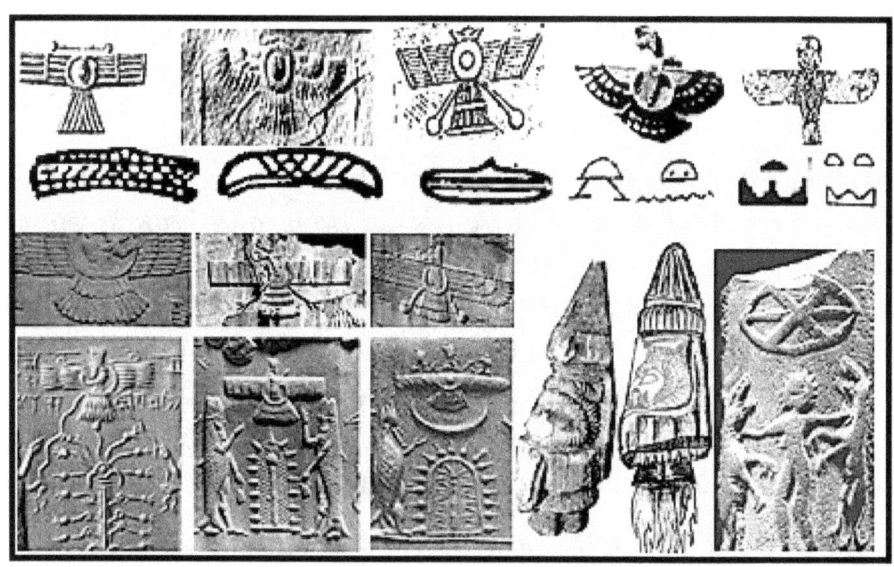

We already know the science of flying is practical and these details are telling us when the first flights were accomplished. The ancient texts also indicate the near planets were colonized. One in particular was called Rahab in the Bible. Today we know it as Venus. If scientist had just read the Biblical and other ancient descriptions of its destruction they would not have to hide so very much physical evidence.

Colonizing Planets

The Bible called inhabited planets "waters" but the details are unmistakable. While the idea of planetary colonization sounds like fantasy and many rejected the idea of this ancient colonization, today we are finding out that colonization was accomplished which supports the Judeo-Christian histories. The saddest description of colonization is associated with the planet known as Rahab.

*Jeremiah 10:13-14- He hath made the Earth by his power, he hath established the world by his wisdom, and hath stretched out the heavens by his discretion. When he uttereth his voice, there is a **multitude of waters/ [inhabited planets]** in the heavens.* It would not make sense that God would sprinkle water in the universe. We can believe that this verse is talking about inhabitable planets sprinkled in the heavens. The ones close to the earth probably became "inhabited".

*Genesis 1:2-7-And the Earth was without form, and void; and the Spirit of **God moved upon the face of the waters/ [inhabited planets]**. -- And God said, Let there be a firmament in the midst of the waters/ [planets], and let it divide the waters from the waters. And God made the barrier, and divided the waters/ [inhabited planets] which were **under** the barrier from the waters/ [inhabited planets] which were **above** the barrier: and it was so. And God called the barrier Heaven.* Water on either side of the heavens does not sound like regular water because it isn't and the water canopy theory doesn't hold water so to speak. The firmament/barrier discussed here

148

seems to be the asteroid separation between the two major inhabitable areas, the terrestrial planets like Earth and Mars; and the Jovian Planets like Jupiter and Saturn.

Psalm 148: 2-4-Praise ye him, sun and moon: praise him, all ye stars of light. Praise him, ye heavens of heavens, and ye waters/ [inhabited planets] that be above the heavens. This is not talking about the old "water canopy" that would have come down during Noah's worldwide flood, because this was thousands of years later and the barrier was still there. David was talking about what the universe was like during his time and the whole water canopy theory doesn't fit. This was talking about inhabitants of "other" planets praising the creator God. Please note the difference between stars/luminaries and waters as well.

Planet Rahab Destruction

While scientists originally made up stories of Venus simply bursting into flames from something they called the "Greenhouse Effect", they knew it was stupid, but it caught on for a while. Soon, they found out that the features on Venus would not support the crazy notion, but they didn't want to be embarrassed so many simply went quiet as new evidence shows the truer image that was supported by Judeo-Christian documentation. We find, according to much evidence and confirmed by many ancient texts, something near Venus destroyed the planet and sent thousands and thousands of meteors towards Earth as Venus caught fire. Not only are there hundreds of ancient texts confirming this event, but also huge amounts of physical evidence to confirm it. Venus's moon essentially exploded and sent the planet into its present fireball state and thousands and thousands of moon pieces hit Venus and many hit the earth. The Bible never references Venus as Venus, but instead, we read about the "Morning Star" or "planet Rahab" and how it was destroyed during the very ancient war. The books of "Psalms", "Isaiah", and "Job" confirm what Genesis identified as only violence.

The Bible indicated the **planet Rahab** was destroyed. As I mentioned before, the Bible tells us God destroyed Rahab and caused it to shatter. Earlier scientists indicated Rahab, or Venus, was destroyed eons ago because of something they called the Greenhouse effect. Today we know the event that destroyed Venus was fast and furious to set the planet on its side. Greenhouse gas does not do that. Neither science not the Bible can separately answer us adequately, but together, the answer becomes clear. Let's look at Judeo-Christian works.

Psalm 89:5-11 And the heavens will praise your wonders, O LORD; --For who in the heavens can be compared to the LORD?-- You have broken Rahab in pieces, as one who is slain; you have scattered your enemies with Your mighty arm. The heavens are yours.

Rahab [this Hebrew word means a "place of vanity" or "vain planet"] was in the heavens, torn to pieces, affected by some type of war, and its destruction showed that God was master of the heavens. The only thing that remotely is described by this verse is the explosion and destruction on Venus. Some religious groups try to indicate that Rahab could be interpreted as "Egypt the vain city", but it makes no sense. Destroying Egypt does not give you back "the heavens" and it was never "shattered into pieces" like Venus [or the Venusian moon]. Some texts add in Satan to the mix so let's see how he figures in.

Isaiah 51:9-10 Awake, awake, put on strength, O arm of the LORD! Awake as in the ancient days, in the generations of old. Are you not the arm that cut Rahab apart, and wounded the serpent?

The serpent, we will see is discussing Satan's army that must have been stationed on Venus. The time period of the

150

destruction by this verse was in the generation of old, during the Pleistocene well before the expansion of Egyptian power, so don't get this confused.

Job 26:12- *"The boastful Angel [Satan] and his followers rebelled. Yahweh destroyed their dwelling places. He divideth the sea with his power, and by his discretion he smashed Rahab. It was reduced to stones of fire."*

In this verse the boastful angel was again a description of Satan. His followers were giants and this verse says their dwelling place, Venus, was destroyed. If you didn't understand the first time he said is a second time and identified the place as Rahab, Venus. Finally the remark about stones of fire, or meteors, would not mean anything unless the stones were seen by people on Earth as the East Coast of the United States was set ablaze by ½ million meteors at the end of the Pleistocene Age and the Carolina Bays were created.

Enoch 85:1-4 *-A single star fell from heaven- raised up and fed among the cows--I saw many stars which descended and projected themselves from heaven to where the first star was.* Some claim this and similar verses are figurative and depict Satan being thrown from heaven, but sometimes people simply write what they believe. In this case the single star is Venus and the many stars would have been the ½ million meteors from its destruction as we will see soon.

India- *"Atlanteans in Valiaxi flying ships"* and *"Indians in Vimana flying ships"* battle on Earth and Moon as recorded in the "Ramayana".

Maharishi Bharadvaya- In this work there are direct indications of gigantic battles in heaven.

Babylon-In the "Epic of Etana" we read, *"Etana looked down and saw the Earth had become like a hill and the sea a well and so they flew for an hour and Etana looked down and the*

151

Earth was like a grinding stone and the sea like a pot. After the third hour the Earth was only a speck of dust and the sea no longer seen" [The ship, of course, was going into outer space.]

China-Methodology of how to send a detachment of men onto any planet was described in ancient documents from Lhasa. These documents were found fairly recently and have been only partially deciphered. The remaining information is being deciphered as we speak, so we may find out more about the space war in the near future.

Greek History-From Greek legends talking about battles between the gods we are told the following: *"Hot vapor lapped the titans, flames unspeakable rose bright to the upper air [outer space], lightning blinded their eyes."* [Apparently lightning weapons were used in outer space]

***Zadspram- Iranian Bible**-And God said, "I will smite thee, Satan, and the creatures [regenerated dinosaurs] which thou thinkest have produced fame for thee. I will destroy everything about them" And Satan having darkness with himself, he brought it into the sky and left the sky so to gloom that the internal deficiency in the sky extended as much as one-third over the star station --since there was an animation of the Earth through the shattering.*

In this verse, Satan brought his darkness into the sky [possibly to the planet Venus]. Let's say a reasonable star station would be on Venus [our closest planet that had air at that time]. The remark about the "extension of 1/3 of Venus" may be referring to just how many of Satan's people lived on the planet. We are finally told that "when it shattered, the Earth was animated". This would have been the massive meteor storm, massive axis shift, months of storms, tidal-waves, volcanoes, and death. Ezekiel gets into the act and describes the event again.

Ezekiel 28:14- *the anointed cherub who covers [Satan again], you <u>were</u> on the holy mountain of God, and you walked back and forth in the <u>midst of fiery stones</u>,"*

Satan was on the holy mountain of God. We can interpret that as Heaven or somewhere close to God, but then comes the strange verse. Satan walked among the <u>fiery stones or meteors</u>. It is not inappropriate to believe that Satan, the serpent, was involved in whatever was happening at that time. The most likely thing that happened, some 11 thousand years ago, was the explosion on Venus. Some of Satan's soldiers could have been living there at the time along with many colonists, but no one was living there afterwards. This destruction brought on a flood and much more during the time of Noah. He had been minding his own business one minute and God told him to find a wife and have kids because the world was going to be destroyed. Whatever happened to Rahab happened in the very ancient times and before the worldwide flood of Noah. It was torn to pieces and its destruction affected the serpent. It is not inappropriate to believe that Satan, the serpent, was involved in whatever was happening at that time. The most likely thing that happened, some 11 thousand years ago, was the explosion on Venus. Some of Satan's soldiers could have been living there at the time, but no one was living there afterwards.

Argon-The main curiosity, however, found by the Magellan probe was that the atmosphere contains high levels of the isotopes of argon, neon and noble gases. These high concentrations of noble gases could only mean that the <u>current atmosphere of Venus is extremely young</u> [possibly only 11 thousand years ago], because noble gases don't combine with other materials and escape easily into space; even with a thick atmosphere.

Earth Axis Shifted and Island Nations Sank

Besides the war, cannibalism, meteors and fire, we will also find out that the earth's axis shifted <u>and Island Nations sank in the Ocean</u>. We find it all started during the time of Jared, Noah's Great-Great Grandfather. According to Judeo-Christian documents, before the end of Jared's life, the Earth had shifted on its rotational axis which caused island civilizations to be destroyed. Like the other events, the time can be determined to have been 10 to 12 thousand years ago, which coincides with the Young Dryas and Holocene Boundary, the next to the last major destruction period of the world. Just imagine how horrible it was. Even with all this, people still did not worship the creator God. Instead, they worshipped the Nephilim, so the real God did the only thing he thought was reasonable.

Something struck Venus pushing it closer to the sun, almost breaking it in half and, completely ruining the hemostasis, slowing down the orbit to an almost stopped condition, causing the atmosphere to build until the pressure level hit unbelievable levels, structures melted, and rivers died up immediately leaving well defined images of their once great streams. The image following shows the crack in the planet that almost split the planet in two.

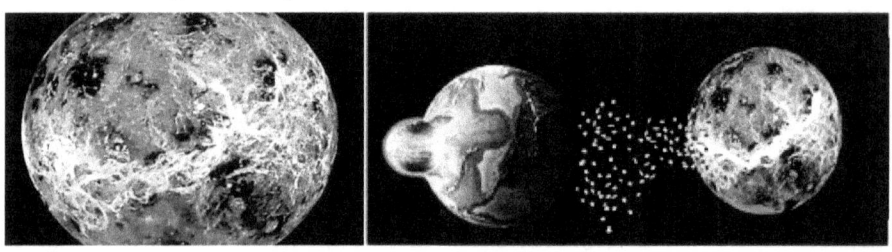

We may even know where it hit as the split expands perpendicularly at the location shown above right. "X" marks the spot. [Above left] If this planetoid struck at a angle oposing the rotation of the planet, it would slow down the rotation significantly as the energy was absorbed. One thing that we know is that most of the massive craters on Venus are concentrated along its Equator suggesting whatever hit Venus was rotating around the planet. One possibility is that the possible destroyer of Venus was its own moon. We know there is a massive electrical difference between the Earth and Venus. Whenever planets come close to one another, it established something called a Plasma, which is sort of like and electrical discharge path without a wire. We use these plasma discharges all the time in Xenon strobes and Fluorescent light bulbs, but when we are talking about an entire planet discharging, it is pretty nasty. One reason to talk about plasmas is that the Soho Satellite registered a substantial, 45 million kilometer long, plasma "electrical discharge string" emanating from Venus and going towards our planet. The image below shows the satellite position when it captured this unbelievable emanation. If the energy was discharged to a planet it might be bad, but if a moon got in the way, the entire structure could be blasted into thousands and thousands of meteors. If the meteors struck Venus at an angle opposite the rotation, it would slow down as we see today. The end of a livable Venus may have looked something like that shown above right.

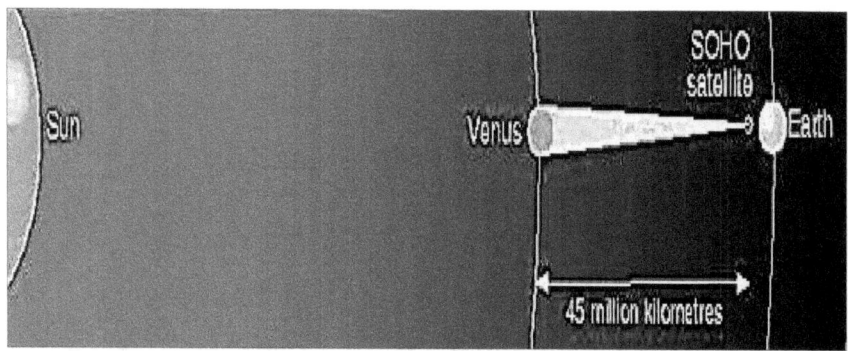

While the plasma string is not visible today, it apparently was visible during ancient times. Here are just a few of the descriptions of Venus, possibly before it turned into a dead planet.

Greek legend, *"A blazing star almost destroyed the world with fire before it became Venus."* Although it is difficult to interpret, I believe they are talking about Venus seeming to have fire coming from it like a plasma blast.

South American Venusian Tail-The people of South America remembered and wrote about it. These Inca legends tell the story. The Inca called Venus the *"Wavy haired planet"*; [This also seems difficult to interpret. Could wavy hair be visible plasma shooting from its surface during a time when the pre-Inca were around?]

Central American Venusian Tail-The people of Central America remembered and wrote about it. This is from one of the Aztec legends. The Aztecs called Venus *"the Star that smoked"* and said that *it once passed by the world blazing and killing many people.* [The "Smoking" sounds like plasma and killing sounds like meteors hitting and setting fires. The meteors came from the "smoking" planet.]

Black foot Indian Venusian Tail-The People of North America remembered and wrote about it. Let's see what the Blackfoot had to say. According to their traditions, *"The morning star* [Venus] *put on a scarlet cloak and appeared before a woman on Earth that he loved. She went into the sky with him, but was warned never to look back. She did, of course, and was ordered to return to Earth."* [The red cloak sounds like a red plasma streamer attached to the planet. The return was a mess if we believe the other histories.]

Ute Indian Venusian Meteors-The Ute Indians tell us the same thing in their verbal history. *"The sun was slivered into a thousand fragments, which fell to Earth causing a general fire. Then Ta-wats fled before the destruction he had wrought. All were consumed; until at last, swollen with heat, the eyes of the god burst and tears gushed forth in a flood which spread over the Earth and extinguished the fire."* [OK; this one is more

156

about the massive meteor shower, but the Ute at least looked in the sky before the Great worldwide flood of the end of the Pleistocene.]

Egyptian Venusian Meteors-In Egypt, the event was known and written about. Sonchie, the high priest, told Solon, a Greek historian, about events before the flood. He wrote, *"Many are the destructions of mankind that have been and shall be. The greatest are by fire and water. During long intervals there are deviations of the bodies that move around the Earth in the heavens and the consequence is widespread destruction by fire of things on the Earth."* [The fires must have been everywhere when the Venusian moon split apart. The comment that it was one of the "Normal bodies that moved around in the Earth sky" limits the body to one of the close planets. Of course, the closest is Venus.]

Sumerian Venusian Tail-The Sumerians made record of the blazing tail of Venus. Their goddess named Inanna was associated with Venus and the information is the same as recorded by all the rest. *"To the queen of the heavens Inanna [Venus], to her who filled the sky with her pure blaze. The luminations are as bright as the sun. Who initiated the flood-storm? You roared in the heavens and Earth. You smote the flesh of the people. She [Inanna/Venus] who causes the heavens to rumble. She who shakes the Earthquake. She cried toward heaven and Earth, "My hair will whirl in heaven for you." You flash like lightning over the highlands. You throw firebrands across the Earth. You split apart the mountains."* The blaze of Venus filled the sky with luminations as bright as the sun as if its hair whirled in the heavens is a pretty picturesque way of describing this plasma thing. The rest describes the horrors of the Meteors and the eventual shaking of the entire earth and the eventual FLOOD STORM.—Man; am I good at this interpretation, or what?]

Assyrian Venusian Tail-Assyrian literature tells the same story. This time the goddess is named Ishtar, but it is the same. *"To the pure flame that fills the heaven, who shines like the sun 'Ishtar" [Venus]—"I ran battle down like flames in the fighting. I make heaven and Earth shake. I trample the Earth. I destroy what remains of the inhabited world"*. Flames in the heavens are addressing the Plasma, but the rest is telling of horrors on the Earth.

Phoenician Venusian Tail-Phoenician texts describe the event, but this time the goddess is Astarte, the Phoenician version of Ishtar. *"See, Astarte" [Venus], she descends into a pool as a fiery falling star"*. This is a beautiful description for a terrible disaster preceded by a plasma tail.

Indian Venusian Plasma-The Indian writers also informed us of this terrible calamity. The people remembered Venus sweeping away the stars. Indian literature states the following, *"Her [Venus's] anger grew so terrible that she transformed herself, grew smaller and black. On a blind rampage she was killing everything and everyone in sight. Her hair is wild, her eyes red. The world trembles and cracks under her tread. Her dark hair flies in the sky* [meteors] *sweeping away the sun and stars."* Again we read about the comet-like tail and so many meteors that the sky is darkened.

Chinese Venusian Plasma-The Far East writers also informed us of this terrible calamity. The people remembered Venus sending down a huge meteor shower. The Chinese writers said the same thing, *"There was a time when a planet [Venus] approached close to the Earth, causing great showers of stones."* Not too many of the planets could have come close to earth, but the plasma tail might have come close and the meteor storm that followed caused the Carolina Bays.

Just imagine everyone running around, yelling and screaming in terror before going to the next section.

Greek History- *"Those innumerable souls, they fall from planet to planet and, in the abyss of space, lament the home they have forgotten."*

Dating the Destruction with Science

Nobel Gas Dating-For this scientists use Argon. A main curiosity found by the Magellan probe was that the atmosphere of Venus contains high levels of the isotopes of argon, neon and noble gases. These high concentrations of noble gases could only mean that the current atmosphere of Venus is extremely young [on the order of thousands of years--say 10 thousand] because noble gases don't combine with other materials and escape easily into space; even with a thick atmosphere.

Evidence of Life still Lingers

Something Making Sulphur-Dioxide-All I can say is three things- Sulphur-Dioxide, Carbon-monoxide, and Ozone. Using data from the Russian Venera space missions and also the US Pioneer Venus and Magellan probes, researchers studying the high concentration of water droplets in the Venusian clouds found hydrogen sulphide and Sulphur dioxide. These two gases react with each other, so they should not be seen in the same place unless something is producing them.

Something using Carbon Monoxide-Despite solar radiation and lightning, the atmosphere contains hardly any carbon monoxide. This suggests something is removing the gas.

Ozone Indicates Current Life -One belief is that bugs living in the Venusian clouds could be combining Sulphur dioxide with carbon monoxide and possibly hydrogen sulphide or carbonyl sulphide in a metabolism similar to that of some early Earth bugs. The Venus Express [2008 to 2012] found Ozone in 2011. This doesn't make sense in that it has only

159

previously been detected in the atmospheres of Earth and Mars. On Earth, it is of fundamental importance to life because it absorbs much of the Sun's harmful ultraviolet rays. Not only that, it is thought to have been generated by life itself in the first place.

Worm Thing- Unbelievably researchers found a live small worm that is somehow still living on the planet. This strongly suggests it has not been long since many animals lived on the planet.The Russian Veruna landed and took a number of pictures before it was finally destroyed by the intense heat, but they seemed to have photographed something strange. Images up until about 90 minutes they just saw the ground, but look at the image taken 100 minutes after landing. Something came out of the ground and disappeared in the 113 minute image.

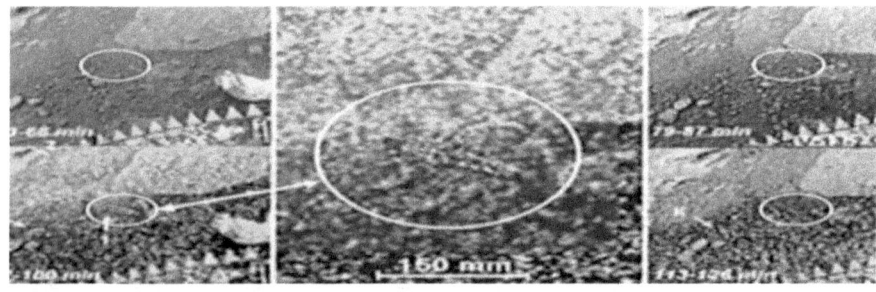

Argon Gas Dates the End of Civilization -The main curiosity, however, found by the Magellan probe was that the atmosphere contains high levels of the isotopes of argon, neon and noble gases. These high concentrations of noble gases could only mean that the current atmosphere of Venus is extremely young, because noble gases don't combine with other materials and escape easily into space; even with a thick atmosphere. Whatever killed most of the life happened not very long ago.

Venusian Remains

All these help us determine that the Biblical stories of Rahab make sense and besides these, there are the remains of

160

civilization and war. On Venus we have a problem in that the massive pressures and heat on the surface today have crushed and melted houses and structures, but we do find what appears to be the remains of a massive war. We have found bomb blasts, huge waterways, roads, and geometrically shaped buildings which all point to colonization. As Venus was destroyed, over a million meteors headed towards Earth.

Buildings-It seems that every few months someone finds another geometrical shaped, building or what appears to be buildings that once must have peppered the landscape before the destruction. Certainly, these are all melted and ruined today, but in some images you can make out the previous grandeur. The high walled U-shaped building is followed by an "H" shaped structure and 2 square buildings. The last on the top row certainly is not a natural structure.

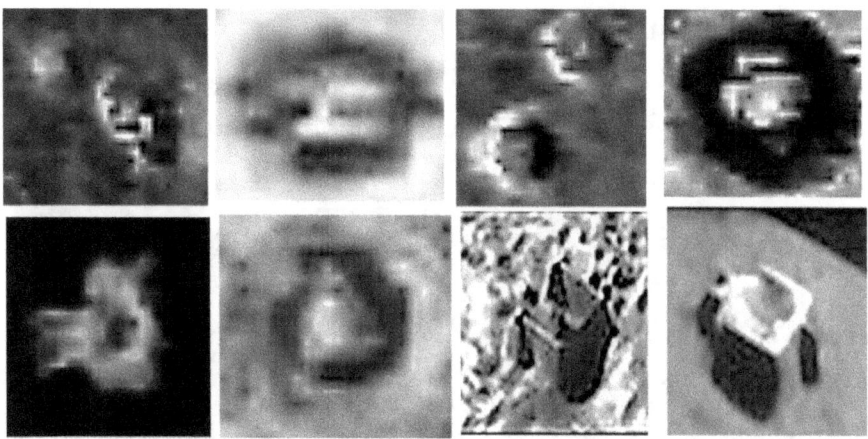

The second row above has another geometric shaped structure followed by a triangular building and 2 square high walled structures. The last 2 have been enhanced to show structure. The top row of the following group shows a geometric complex, massive high walls and then two square buildings that are adjoined. The 4th image shows walls and a square structure and the last on the top row looks like it has a tower of some kind.

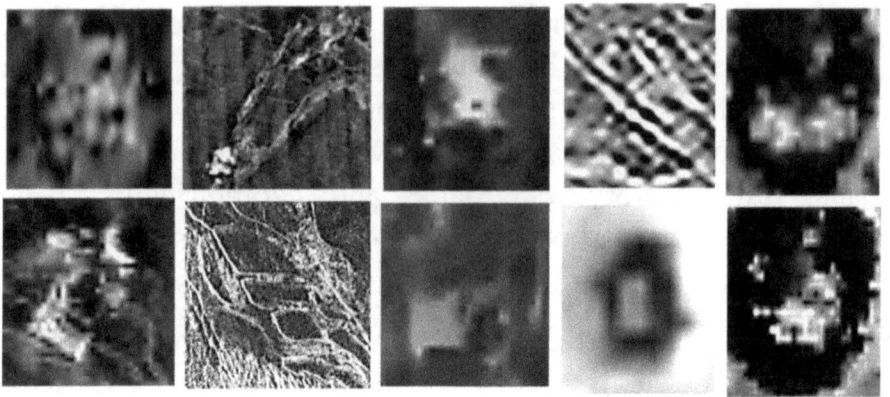

The second row above starts with a multi-component grouping of buildings followed by what could be a three stepped set of square buildings. The three more structures with squared sides end this group. Please understand the detail is substantially less clear through the thick atmosphere than images taken of the Martian and lunar surfaces but we can still see suspicious items that look man-made.

Roads-Over the tops of all types of hills and valleys, Venusian roads seems to go on forever which shows a reasonable probability that someone used those roads.

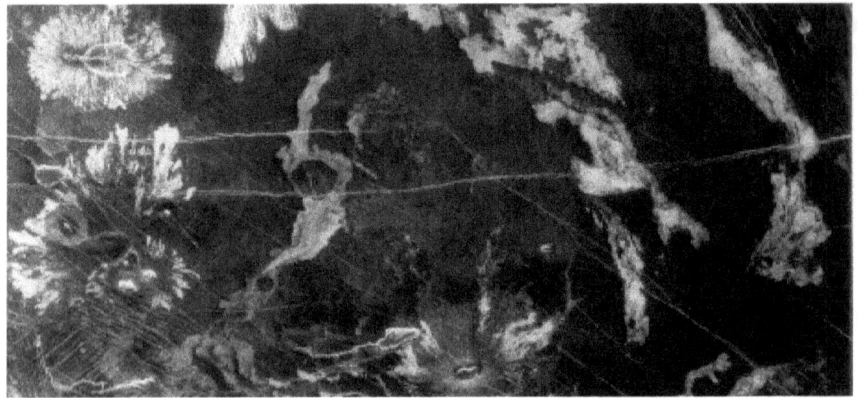

Below are a couple more similar roadways. In the old days these probably went somewhere, but today they are just endless anomaly.

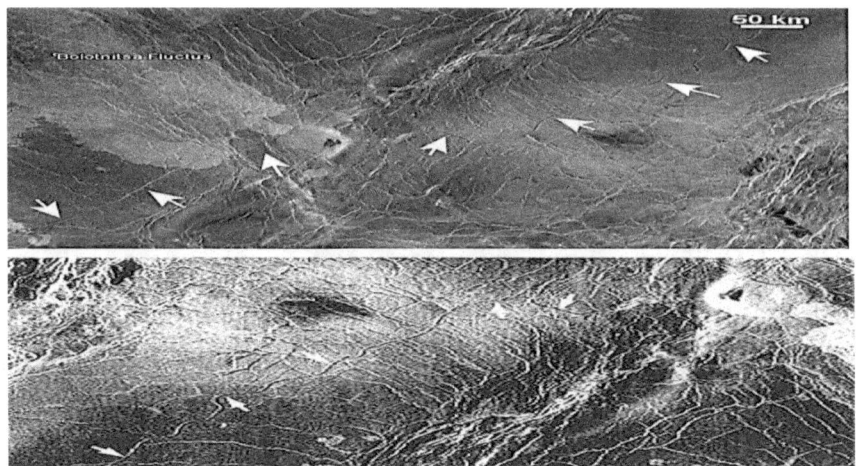

Waterways- We can certainly see that Venus once was filled with beautiful waterways and massive rivers. Today they are completely dry, but it happened such a short time ago that the details of the rivers are still clear. The waterway on the right appears to have had some type of commerce before the destruction of Rahab.

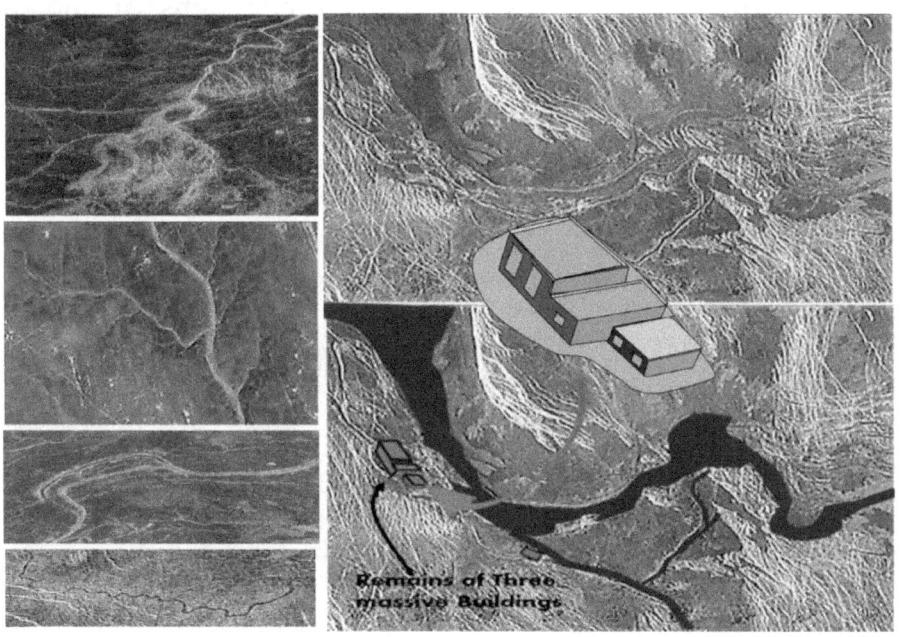

Let me leave this alone, after I just show a few of the thousands of images collected to date as colonization must have been widespread around this time. While we find the

163

l;arge number of cities or industrial areas on Mars, while the lunar surface is filled with smaller building groupings.

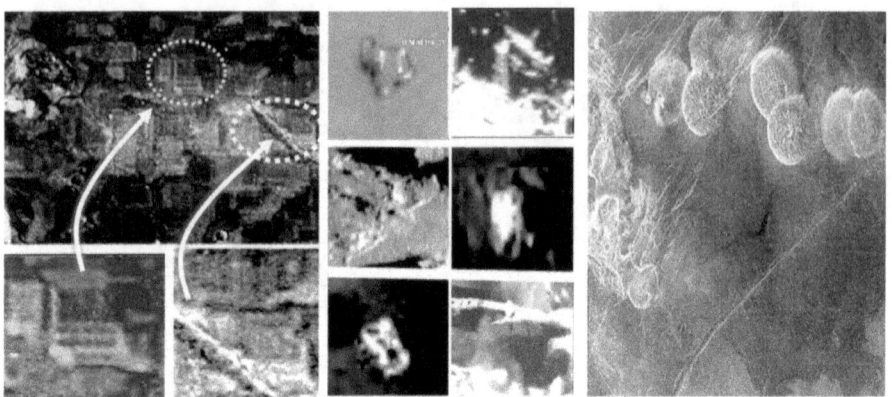

Martian Remains   Moon Remains   Venus Remains

Mars Remains-The first group shows just one of many industrial sites or what used to be. Nortice the building [first image bottom row] with glassed windows. We also have found an abundance of water, trees, pyramidic shapes, melted areas in towns. Roadways and many other signs of ancient colonization.

Lunar Remains- The middle sections shows a tiny grouping of artifacts from colonization on our moon. These include walled in industrial areas, lighted becons, many towers, geometrically shaped buildings and artifacts that appear to be crashed aircraft as well as construction equipment.

Venus Blast Area- On the end is another image from venus showing 7 identical blast areas near one of the roadways. These appear to be bomb craters because all craters had almost identical impacts. If we want to see what massive meteoric crater blast fields are like we need to look on the East Coast of the United States.

Carolina Bays

As Venus turned into a blistering, high pressurized hell, the east coast of the United States was pelted with many objects. There are still an estimated **500 thousand** meteorite indentions called "Carolina Bays", which mark this incredible event in history. One hundred and forty thousand of these holes have diameters of over 500 feet. Just think about how afraid the people of that time were as they essentially saw the sky fall all around them—and this was before the Earth axis shifter and the rains came and the tidal waves covered the mountaintops and massive extinction of almost all life. The picture below right shows the major areas where these objects have been found in the United States. The globe on the left shows the various equators the Earth has had according to the Hot Spot trail of the Hawaiian Islands. Notice most of the time that mankind has been on the earth, the equator has been along the plane parallel to the US east coast. These generally date around the same time. The evidence shows that the Venusian moon met its end at the same time that these 500 thousand holes appeared. Some of these indentions are very large and have diameters that are thousands of feet across. So it wasn't just a little meteorite storm.

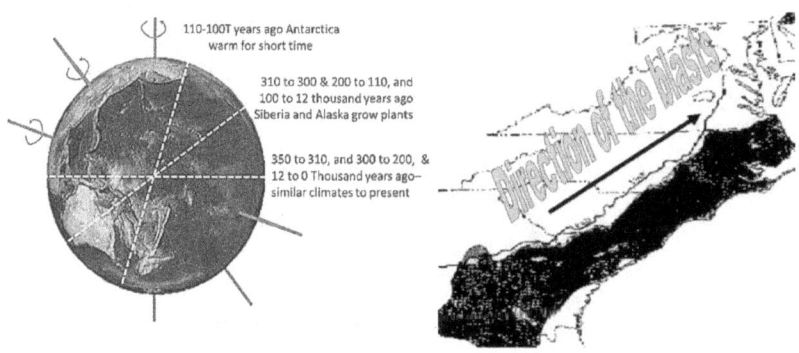

110-100T years ago Antarctica warm for short time

310 to 300 & 200 to 110, and 100 to 12 thousand years ago Siberia and Alaska grow plants

350 to 310, and 300 to 200, & 12 to 0 Thousand years ago— similar climates to present

Direction of the blast

The direction of the blast shows the equator before the meteor storm disruption. The Carolina Bay incident was a huge onslaught of meteors striking the Earth, which caused holes everywhere. Don't just take my word on this. Next is a picture showing the quantity of these things in a small area of South Carolina. Just think about how it must have been that long time ago if you happened to live near South Carolina and literally hundreds of thousands of meteorites blasted the entire area over a period of perhaps 6 months. Most of your friends would have been killed and any civilization that was built up would have been in shambles. That is what the world would have been like, but the worst wasn't over—and for those who think that the Carolina Bays look like a multitude of sink holes caused by eroding caverns in an underground aquifer, think again. The underground area below these "indentions" is not a limestone honeycomb and sinkholes could not be the cause of half a million craters. These came from meteors and the meteors all came 10 to 12 thousand years ago.

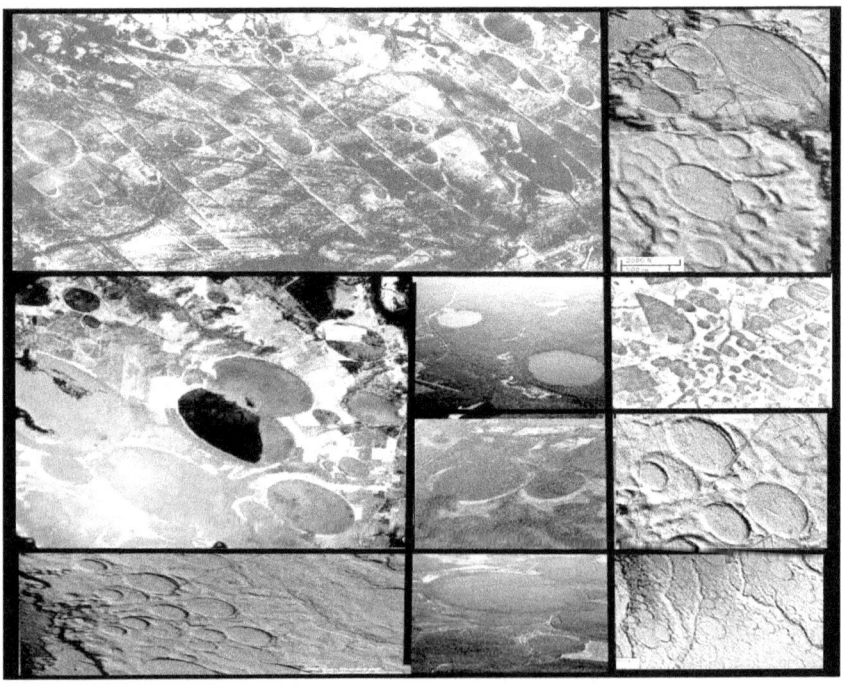

166

We can assume this path was along the equatorial path of the Earth at that time The image below left shows how an equator along the eastern coastline of USA also would travel around to the south west portion of Australia and it would have Siberia in the tundra region allowing hundreds of thousands of Mammoths to graze there before being quick frozen as the earth axis shifted soon after this time. To the right is a satellite image of some of the crater lakes in the southwestern part of Western Australia along the same ancient equatorial path as parts of the Venus moon pelted the other side of the world.

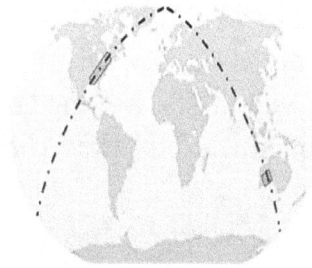 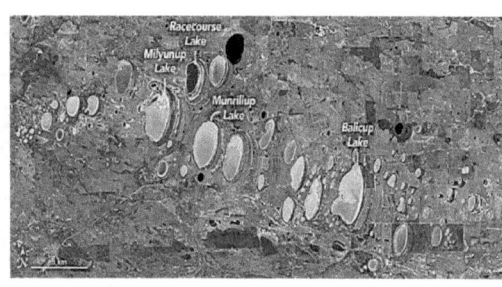

The bombardment of the Venusian moon particles didn't simply happen and afterwards things went back to normal on earth. The effects of the bombardment, evidently, set up a chain of events including pushing volcanic debris in the sky and initiating a heating cycle on the earth. Inhabitants were initially unaware of how catastrophic this would be, but soon their fate was known.

Pleistocene Extinction

Within a thousand years of the destruction of Rahab, the whole Earth would be flooded, but that was only a small part of the destruction. Judeo-Christian documentation tells us Noah built a huge floating hotel for animals just before the whole earth began to shake, the sun darkened, and the foundations of the world raged as the foundations of the world were broken up. Then the earth axis shifted, forcing the mammoth herds into the Arctic Circle. Continuous rain, the polar Ice caps, melting and reforming, and unbelievable tidal waves would cover the world with water and deposit almost mile high mud-piles filled with the carcasses of millions of fish in Karoo Africa. It would take scientists a long time before they would admit the Earth Axis shifted as they didn't read the Bible and the associated histories.

Jasher 6:11-14 And on that day, the Lord <u>caused the whole Earth to shake, and the sun darkened, and the foundations of the world raged, and the whole Earth was moved violently, and the lightning flashed, and the thunder roared, and all the fountains in the Earth were broken up</u>, such as was not known to the inhabitants before; and God did this mighty act, in order to terrify the sons of men, that there might be no more evil upon earth. And all the fountains of the deep were broken up, and the windows of heaven were opened, and the rain was upon the Earth forty days and forty nights.

"Harris Papyrus"-A cosmic cataclysm [meteors] of fire and water followed with the <u>south becoming the north and the Earth turning around</u>

Emerald Tablets-Called he the three mighty messengers. He gave the commands that shattered the world. Called he on the

168

powers of the seven lords. Wielded; the power <u>changed the</u> <u>Earth's balance</u>. Down sank Atlantis beneath the dark waves. [This one also talks about the major Island of Commerce sinking at that time.]

Another Egyptian Tradition-The king of Egypt saw, in his sleep, the <u>whole Earth turned over</u>. This, it was said, foretold as destruction by flood.

Enoch 64:1. *In those days Noah saw that the earth became <u>inclined</u>, and that destruction approached. And Noah cried with a bitter voice, --what is transacting upon earth; for the <u>earth labors, and is violently shaken</u>. Surely I shall perish with it. After this there <u>was a great perturbation on earth</u>, and a voice was heard from heaven. .— Their <u>seed shall be</u> <u>backward in their prolific soil</u>; In those days the fruits of the earth shall be late, and not flourish in their season; and <u>in</u> <u>their season the fruits of the trees shall be withholden</u>. The <u>moon shall change its laws, and not be seen at its proper</u> <u>period. Those shall not appear in their season</u>.*

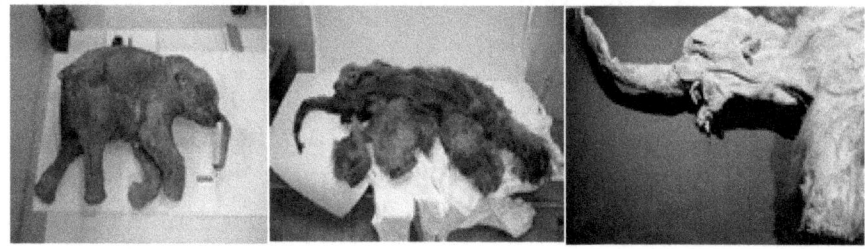

At one time Siberia had been a paradise and the massive animals were eating flowers. As the earth shifted, they were quick frozen with flowers still in their mouths. Here is just a tiny sampling of baby Mammoths frozen with their mothers.

Millions of mammoth tusks, perhaps more, are still locked in Siberia's permafrost and about that many have been recovered for the lucrative Ivory trade. Nearly 90 percent of all mammoth tusks hauled out of Siberia, estimated at more than 80 tons a year, ended up in China, with tusks that spiral to

169

more than 13 feet. While it is believed the tusks were first harvested during the late Roman era, by the early 1900s, massive warehouses for the ivory could be found all over as shown below. The Pleistocene Extinction is making many wealthy.

If scientists had read the Judeo-Christian texts, they would have easily determined what had happened and why. Some may have been afraid that the Bible would have to be believed so the massive evidence of the worldwide flood during the Pleistocene Extinction was simply not provided as something teachers should bring up to our children. I just have to tell you it seems that scientists always want to do it the hard way. First someone comes up with a stupid theory, then after decades of testing it is decided that someone should look in the Judeo-Christian documents. Once that is done they can completely change the original theory and allow people to see the things held as anomalous for years. This is the case of the next scientific discovery as well this one is commonly known as the Bharata War. While we don't know exactly when it started we know with a good amount of certainty it ended about 3100BC.

Science and the Bharata War

Judeo-Christian texts confirm the Bharata War sometimes called the Tower of Babel War, but many historians are still reluctant to talk about it. Similar, in some ways to the massive was before the end of the Pleistocene. Like the previous worldwide war, this one also used horrible weapons and even nuclear type weapons, but this time, the war was located only on Earth. One thing that was very similar is that about half of all the major mutations of humans happened during this great event.

The great "Tower of Babel" in Baalbek, Lebanon had been destroyed; huge numbers of cities were decimated; massive walls had been constructed and had melted under intense heat from battles; and death was everywhere. The book of "Jasher" tells us 1/3 of the entire population at that time lost their lives in the Bharata War and another 1/3 were thrust back into living like animals. Finally, the Great War ended in 3100BC. We won't go into the horrors of this massive war too much, but we do know the Maya started a brand new calendar with the end of the war in 3100BC as the beginning of a new world. Yes, the calendar did end in December of 2012, but that has little to do with the story. The Dravidians in India claimed a brand new "Age of Kali" starting in 3100BC. The Egyptians claimed the "beginning of a New Age" called Zep-Tepi [New Beginning]. This happened in 3100 BC and the Brazilian Mongulala people ended their "Blood Age" and started the new "Present Age" in 3100 BC. The Bible indicates Peleg [which means World Divided] was ruler of the Adamic line but then, around 3100BC, a new ruler took his place named Reu [which means Friendly or Joined World]. The chart following shows these representative nations in a rough timeline.

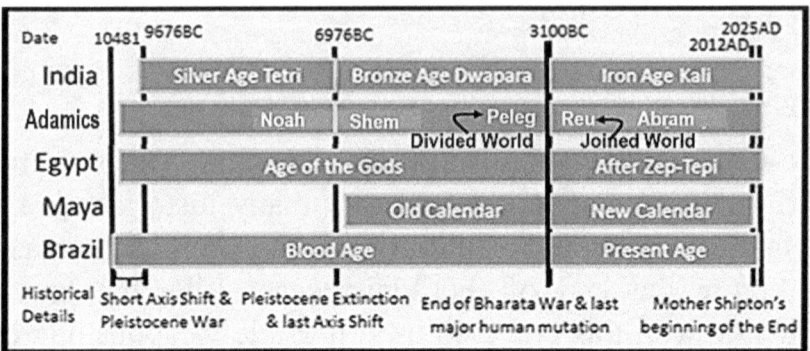

Some of you know more about this War than you know in that the Canaanite named King Nimrod built a huge citadel and Tower at what would be one of the strongholds of the Phoenicians in Baalbek, Lebanon. One of the names given the site was the Tower of Babel. During the war, the citadel was destroyed, but King Nimrod escaped and began ruling Ur or what is now Iraq. The following collage shows some of the remaining blocks used in the construction of the huge citadel and tower. The tiny circles are around people standing on or around these huge stones. Many believe only the giant Anak people would have been the only ones able to move these 1000 ton massive stones and place them into position. Some have tried to move the stones with modern machinery and it has proven to be almost impossible, but dozens of these massive building blocks attest to people with unbelievable strength or abilities we are just now beginning to understand like levitation and anti-gravity. Please notice something very important in the next collage. The massive, impossible to move stones as shown in the bottom right image are not at the bottom of the structure. They were so easy to place, they were put down as the 4th and 5th layers. Also if you notice the top image; at the top of that image there are some columns. That is

172

where the completely finished stone originally was placed as that is the site of the Citadel and tower. It is as if something flung the stone like it was paper during the war.

While there was destruction everywhere and many societies began living underground during this war, scientists had been trying to find out what happened in a major Indian city called Mohen ja Doro [City of the Dead]. Here we find skelatal remains of people clumped together with no way to leave and pottery melted down to make balls of glass and walls with the blocks melted together. Strangely, the skeletons are still radioactive today. The following collage is a small sampling of the dead caught in the middel of a nuclear war.

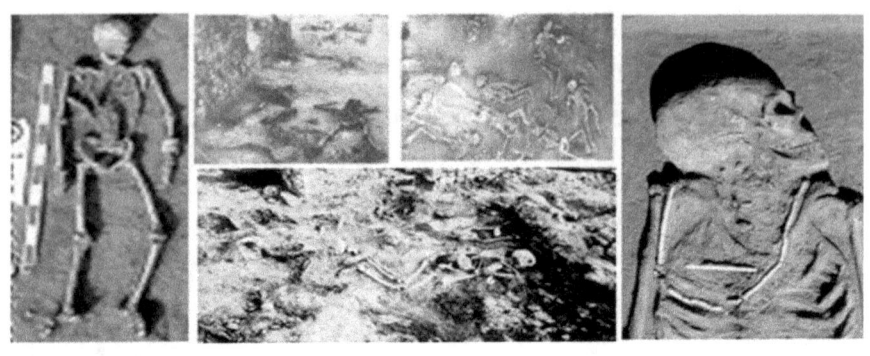

I know you are not reading about this major event in history, but the Judeo-Christian texts are not the only descriptions found. The Indian Mahbharata text describes the wars in great detail, and the underground cities in Turkey, Malta, Brazil, Peru, Guatemala, China, Scotland, France, the United States and just about everywhere attest to a world too afraid to even live outside of an underground hole. The hapotype scientists confirm what they were afraid of.

Haplotype Scientists Were Puzzled

As I mentioned almost half the major DNA mutations of modern man occurred during the Pleistocene War and the other half were mutated during this war as shown from the partial listing of mutation below.

Mutation indicator	Time [x1000 yr]	Races of Men caused by mutation
P	6	Proto Ameridian
R2	6	Aryan
N	6	Russian
O	6	Oriental
L	5	Dravidian/Indian
R1a	5	Slavic
R1b	5	Gaelic
N	5	Scandinavian

A new DNA study indicated that the genetic markers of the early Western European groups were suddenly replaced around 5 thousand years ago. Anthropologists were stumped. This included those involved in the National Geographic Society's Genographic Project, with 700 thousand researchers

174

in 140 countries around the world. By not accepting the massive nuclear war from this time, they simply have no explanations. They can tell there was a dramatic series of events including major migrations from both Western Europe and Eurasia, signs of an unexplained genetic turnover about 5000 years ago, and the huge number of major mutations, but no one wants to say it was a war when many ancient historical references describe it and physical evidence is everywhere. The above ground destruction, the people having to live underground; the radioactive skeletal remains; the entire deserts full of melted sand turned into what is called Libyan glass; and melted blocks on fortifications in Scotland, France, Peru, India and many other places [a sign of intense heat from projectiles]. If they say it was war, the Bible would have to be proven again and the Tower of Babel War would be taught in school where it should be..

As an example-Maternal "H" and the Paternal "R1b" Haplotype both showed up together as the beginning of a group called the Beaker people or Gauls. The map below shows that they were simply localized in mid European area.

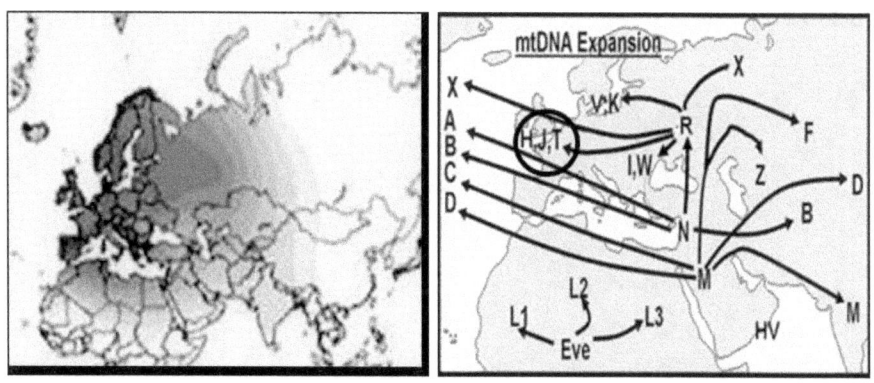

This means five thousand years ago, <u>something changed drastically</u>. Huge mutations showed up in the MtDNA and allowed for the production of the I, J, K, W, H, V, T, and Q Haplotypes almost overnight. On the Y-Chromosome side, the R, P, N, O, L, and N Haplotypes also seems to come from nowhere 5000 years ago. We had begun a new world. As recorded in Babylonian, Egyptian, Chaldean, Jewish histories, people all of a sudden only lived for a hundred years rather than thousands as if DNA mutated. Our brain size began to reduce and around the world we were told people could no longer think clearly, they became more like animals, they no longer could understand the speech of their co-workers, they could no longer see great distances, and that only 10 percent of our brain could be used as our brain began to atrophy from disuse. The mutations had affected use mentally and physically, but some scientists still want to hold up blinders and keep saying , "That is just an anomaly, that is just strange, that could not have been."

Rather than continuing in the Old Testament, I think a transfer to the New Testament is warrented as we find the one called the Messiah or Jesus. The greek word used for the Messiah may be more correctly translated "Joshua", but the question this book is building continues to be, "Is there scientific evidence of Judeo-Christian religious document's truth?" The first truth to look for in th eNew Testament should be was Jesus real?

Forensic Science Was Jesus Real?

This is possibly the most important thing. Did large groups of people go to their death for this Jesus guy? Did the disciples purposely incite people with claims of his god-ness knowing they would be killed in the most horrible ways? Some say the New Testament was "all" made up. They whine, "How can anyone believe a history that is so full of out and out mis-information?" "Jesus didn't even exist!" they proclaim. Here are some of the topics I found were brought up when trying to eliminate Jesus from Science and history. All have been found to be false, but that has not halted continued attempts trying to eliminate Jesus from history.

- There is no proof that Jesus ever existed. Way False
- If he did exist, there is no record that he was a well-known orator. Way False
- There is no evidence that he rose from the dead, in fact contradictions about this even are huge in the Gospel accounts. Way False
- There is no indication Pontius Pilate ever ruled. Way False
- There is no evidence that Crucifixion was used. Way False
- There is no evidence that there was a three hour period of darkness and there could not have been a solar eclipse in the season of his death. False
- There is no account of Jesus having large groups following him while he was alive. False
- The evidence brought forth to support Jesus is quickly determined to be fake. False, False, False

These things just don't make sense. Here are just a few of the hundreds of bits of evidence that have been preserved over the 2000 year period. One might be amazed at how much information is known while commonly "known" people like Plato have almost no true evidence of existence. I'm not saying Plato didn't exist, I'm just saying this stuff is simply amazing.

Scientists and researchers have found all type of information on Jesus including Roman documents, Jewish documents, Jewish Histories, Census details, and many other things showing Jesus was real, crusifixion was the favored death mode including nailing feet to the wood. The famed Shroud of Turin with a photgraphic image on the sheet is pretty amazing. Shroud of Turin and the painting of Jesus by the woman who was cured by Jesus. Have been used to give us a possibly glimse of what this incarnation of God looked like.

The Turin cloth has a wealth of information and confirmation.

Blood Stain Information-Dr. Baima Ballone analyzed samples taken from the cloth. He reported that the blood on the shroud is type AB. This is significant in that type AB is almost nonexistent in rare in Europe, Australia, and the Americas, with an estimated 3.2% of the world's population having this type. That being said, around 18% of Jews have it, so the probability this image came from an ancient Jew is substantial. By the way, in a male, the blood type comes from the mom not the dad, so don't go thinking we have found the blood type of God.

Floral Evidence-The British Broadcast Company (BBC) reported on the 1999 conference of the Missouri Botanical Society: "Professor Avinoam Danin of the Hebrew University of Jerusalem said an examination of pollen traces and floral imprints suggested that they could only have come from plants growing in a restricted area around Jerusalem and would date back to Jesus' time. He said: "*This combination of flowers can be found in only one region of the world. The evidence clearly points to a floral grouping from the area surrounding Jerusalem.*" The pollen grains were collected from the shroud some years ago. His researchers also said a type of pollen from a thistle visible near the shoulder of the man's image on the shroud was believed to have come from the plant that would, most likely, have been used for Jesus' crown of thorns. Wow! The cloth came in contact with the environment of Jerusalem just before being imaged.

The Image Evidence- The method of making the image seems supernatural and has never been found or duplicated ANYWHERE! The image having the wounds that are similar to those described in the Biblical account seem to add credence to the fact that Jesus did die, was wounded, and released some "energy" to deposit the image found.

Evidence Of 1st Century Roman Coins-Enhanced photography techniques and computer analysis was used in

1979 at University of Chicago. It was observed on the right eyelid of the shroud image four letters "UCAI" which formed a crown around the crook of an augur's staff.

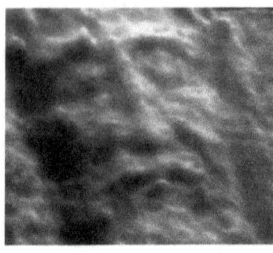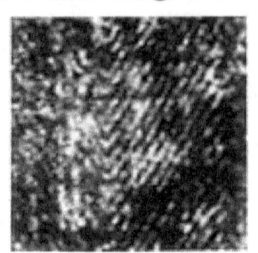

This image corresponds to the symbol on a small coin known as the dilepton lituus struck in 29 AD during the procuratorship of Pontius Pilate (26-36 AD). The coin covering the left eye was later identified as the lepton simpulum, which also was struck around 29 AD. Jews in the 1st Century AD used coins to hold down the eyelids. This person died a little after 29 AD. It's like putting a time stamp on the cloth including the general location.

Burial Certificate Evidence- To make this even more reasonable, in 2009 a researcher in the Vatican secret archives managed to read the burial certificate of Jesus the Nazarene. She had reconstructed it from fragments of Greek, Latin and Hebrew writing imprinted on the shroud. Indications included the Greek words *Isous Nazarennos* [Jesus the Nazarene], and *Tiberiou* [Tiberius], the Roman emperor of the time of Christ's crucifixion.

Josephus [93AD] was a Jewish historian, as such he would have wanted to insure the information about Jesus did not go out, even so, he had to report on a major event." *About this time there lived Jesus, a wise man, if indeed one ought to call him a man. For he was one who performed surprising deeds and was a teacher of such people as accept the truth gladly. He won over many Jews and many of the Greeks. He was the Messiah. And when, upon the accusation of the principal men among us, Pilate had condemned him to a cross, those who* *had first come to love him did not cease. He appeared to them spending a third day restored to life, for the prophets of God had foretold these things and a thousand other marvels about him. And the tribe of the Christians, so called after him, has still to this day not disappeared."* This still adds weight to the considerable body of historical evidence for the existence of Jesus and his death by crucifixion. The account by Josephus of the death of James, brother of Jesus, widely accepted even amongst the most independently minded scholars, is further confirmation that Jesus existed as a historical figure--

Jewish Antiquities- *"The younger Ananus, who had been appointed to the high priesthood, was rash in his temper and unusually daring. He followed the school of the Sadducees, who are indeed more heartless than any of the other Jews, as I have already explained when they sit in judgment. Possessed of such a character, Ananus thought that he had a favorable opportunity because Festus was dead and Albinas was still on the way. And so he convened the judges of the Sanhedrin, and brought before them the brother of Jesus, the one called Christ, whose name was James, and certain others, and accusing them of having transgressed the law delivered them up to be stoned."*

Written Evidence -Cornelius Tacticus, a Roman Historian reported the following in 115AD- *"Consequently, to get rid of the report, Nero fastened the guilt and inflicted the most exquisite tortures on a class hated for their abominations, called Christians by the populace. Christus [Christ], from whom the name had its origin, suffered the extreme penalty during the reign of Tiberius at the hands of one of our procurators, Pontius Pilatus, and a most destructive superstition, thus checked for the moment, again broke out not only in Judea, the first source of the evil, but even in Rome, where all things hideous and shameful from every part of the world find their center and become popular. Accordingly, an* *arrest was first made of all who pleaded guilty; then, upon their information, an immense multitude was convicted, not so much of the crime of firing the city [Rome had been burned 6 days before this record], as of hatred against mankind. Mockery of every sort was added to their deaths. <u>Covered with the skins of beasts, they were torn by dogs</u> and perished, or were **nailed to crosses**, or were <u>doomed to the flames and burnt</u>, to serve as a nightly illumination, when daylight had expired."*

Written Evidence -Origen, a Roman Historian, wrote about the same incident. *"---and with regard to the eclipse in the time of Tiberius Caesar, in whose reign Jesus appears to have been crucified, and the great earthquakes which then took place .. "*

Written Evidence -Suetonius another Roman writer (A.D. 75-160) considered Jesus as a Roman insurgent who stirred up seditions under the reign of Claudius (A.D. 41-54) even after he died.

Written Evidence - the Jewish Talmud, another ancient document written shortly after Jesus' death refers to Jesus having been crucified on the Eve of Passover.

Written Evidence- Pliny the Younger, (61 AD – ca. 112 AD) Pliny served as an imperial magistrate under Trajan (98–117AD). He reported on his actions against the followers of Christ. He asked the Emperor for instructions dealing with Christians and explained that he forced Christians to curse Christ under painful torturous inquisition: *"They were accustomed to meet on a fixed day before dawn and sing responsively a <u>hymn to Christ as to a god</u>, and bound themselves to a solemn oath---"* So Pliny was aware of Jesus as a God. In a later writing he details persecution against Christians: *"Even this practice, however, they had abandoned after the publication of my edict, by which, according to your*

182

orders, I had forbidden political associations. I therefore judged it so much more the necessary to extract the real truth, with the assistance of torture, from two female slaves, who were styled deaconesses: but I could discover nothing more than depraved and excessive superstition. In the meanwhile, the method I have observed towards those who have denounced to me as Christians is this: I interrogated them whether they were Christians; if they confessed it I repeated the question twice again, adding the threat of capital punishment; if they still persevered, I ordered them to be executed. For whatever the nature of their creed might be, I could at least feel no doubt that contumacy and inflexible obstinacy deserved chastisement. There were others possessed of the same folly; but because they were Roman citizens, I signed an order for them to be transferred to Rome." We can see Pliny was no Christian, he tortured them, threatened death, and they refused to deny their faith in Jesus. This is remarkable evidence of Jesus' existence outside of the Bible.

Written Evidence -Sextus Julius Africanus (160 – 240AD) quoted the Greek historian **Thallus** [1st century writer] *"An eclipse of the sun unreasonably, as it seems to me (unreasonably of course, because a solar eclipse could not take place at the time of the full moon, and it was at the season of the Paschal full moon that Christ died.)"* *[Parenthetic statement by Africanus].-* Bible states in reference to the Jesus dying on the cross: *"Now from the sixth hour there was darkness over all the land unto the ninth hour"* *[noon until 3PM]*.

Written Evidence -Phlegon-a Greek historian from 2nd Century AD the odd eclipse as well. Africanus also quoted his writing of an eclipse occurring on or about the day Jesus was crucified. It was clear that Phlegon did not know from his sources about any similar eclipse in previous times.

Written Evidence -Lucian- was a Greek lecturer (115BC) who mocked Christians in his writing, providing evidence that Jesus really did exist: *"He was second only to that one whom they still worship today, the man in Palestine who was crucified because he brought this new form of initiation into the world.---Having convinced themselves that they are immortal and will live forever, the poor wretches despise death and most willingly give themselves to it. Moreover, that first lawgiver of theirs persuaded them that they are all brothers the moment they transgress and deny the Greek gods and begin worshiping that crucified sophist and living by his laws.--"They scorn all possessions without distinction and treat them as community property. They accept such things on faith alone, without any evidence."*

Written Evidence -Suetonius was a 1st century Roman historian and this is what he wrote: *"The emperor Claudius reigned 41 to 54 AD. Suetonius reports his dealings with the eastern Roman Empire, that is, with Greece and Macedonia, and with the Lycians, Rhodians, and Trojans. He then reports that the emperor expelled the Jews from Rome, since they "constantly made disturbances at the instigation of Christ".* Skeptics will point to a different spelling to say "that's not Jesus he's talking about!" but it is obvious that followers of Jesus in the Roman Empire were persecuted by Roman authorities.

Dead Sea Scroll Evidence-One of the most important archaeological finds that actually dates to the time of Jesus may provide evidence of Jesus existence. The parchment documents found were written sometime between 150 B.C. and A.D. 70. In one place, a scroll refers to a "teacher of righteousness." Some say that teacher is Jesus showing the Essenes became some of the very first Christians

Bone Evidence-A limestone box was found along with 800 or so other well preserved bone boxes, but this one was inscribed

184

with "Ya'akov bar Yosef akhui di Yeshua" ("James, son of Joseph, brother of Jesus"). A team of experts from the Geological Survey of Israel examined the box and the inscription under a microscope and found no evidence of modern tools or tampering. Like the rest of the box, the inscription, though wiped clean in parts, had a thin sheen of particulate matter formed on it called a patina. This particular patina shows that it developed in a cave environment and that it is consistent with an age of 2,000 years. Such bone boxes were in use from about 20 B.C. to A.D. 70, when according to Jewish custom the dead were first sealed in caves or rock-cut tombs, then their bones later transferred to a limestone bone box after the body had decayed. It was found that its inscription was in a cursive style used only in the few decades before A.D. 70, when Jerusalem was destroyed by the Romans. Thus, the inscription fits the style used around A.D. 62, when James, Jesus' half-brother, died. It was noted that the ONLY reason for naming the brother was because of the brother's notoriety of the time. By the way; Matthew 13:55-56 tells us the names of Jesus "half" brothers and indicates he had sisters as well. The brothers' names were James, Joseph, Judas, and Simon. The sisters were unnamed.

Ankle Bone Evidence-A 2000 year old ankle bone of a crucified man was excavated in a burial cave at Giv'at ha-Mivtar, northeast of Jerusalem, in 1968. This cave contained five ossuaries or bone boxes. In one of the ossuaries were the bones of a young man who had died in his mid-twenties, crucified at about the same time as Jesus. A 4.5inch nail was still lodged in the heel bone of the man - apparently the people who buried him had been unable to pull it out. There was even a small wedge of wood remaining between the heel bone and the head of the nail, which had been put there by some Roman soldier to hold the nail and his foot firmly in place. According to the ossuary box inscription, his name was 'Yehochanan', which in English is 'John' and both his leg bones had been smashed, something that was done to hasten the death of the crucified man. This adds another level of evidence to probability that Jesus had been nailed to a cross 2000 years ago along with others. The picture shows the actual artifact and how it might have been done.

Healing the Sick

The next this to examine might be Healing the Sick. What the Bible indicates is that somehow people can cure serious illness with something called faith. I hope you have a better understanding about how this miraculous deed can be done by anyone with faith as small as a grain of mustard seed as Jesus liked to say.

Luke 10:1-17-Jesus sends the Seventy out and commands them, "to heal the sick there, and say to them, the kingdom of God has come near to you.--The seventy return with joy, saying, "Lord, even the demons are subject to us in Your name

Acts 3:1-16-"Then Peter said, silver and gold I do not have, but what I do have I give you: In the name of Jesus Christ of Nazareth, rise up and walk. And he took him by the right hand and lifted him up, and immediately his feet and ankle bones received strength. So he, leaping up, stood and walked and entered the temple with them-walking, leaping, and praising God.

Acts 4:16- "What shall we do to these men? For, indeed, a notable miracle has been done through them is evident to all who dwell in Jerusalem, and we cannot deny it"

Acts 5:12-16-"And through the hands of the apostles many signs and wonders were done among the people. And they were all with one accord in Solomon's Porch". --"Also a multitude gathered from the surrounding cities to Jerusalem, bringing sick people and those who were tormented by unclean spirits, and they were all healed"

Acts 9:33-35 -Aeneas is healed. "*And there he found a certain man named Aeneas, which had kept his bed eight years, and was sick of the palsy. And Peter said unto him, Aeneas, Jesus Christ makes thee whole: arise, and make thy bed. And he arose immediately.*

Acts 6:5-8-"*And Stephen, full of faith and power, did great wonders and miracles among the people*".

Acts 8:6-7-"*And the people with one accord gave heed unto those things which Philip spake, hearing and seeing the miracles which he did. For unclean spirits, crying with a loud voice, came out of many who were possessed; and many who were paralyzed and lame were healed*".

Acts 9:17-18"-*And Ananias went his way and entered the house; and laying his hands on him he said, "Brother Saul, the Lord Jesus, who appeared to you on the road as you came, has sent me that you may receive your sight and be filled with the Holy Spirit. Immediately there fell from his eyes something like scales, and he received his sight at once; and he arose and was baptized.*

Acts 14:8-10- "*And in Lystra a certain man without strength in his feet was sitting, a cripple from his mother's womb, who had never walked. This man heard Paul speaking. Paul, observing him intently and seeing that he had faith to be healed, said with a loud voice, Stand up straight on your feet! And he leaped and walked*".

Acts 13:8-11- "*But Elymas the sorcerer withstood them, seeking to turn the proconsul away from the faith. Then Saul, who also is called Paul, filled with the Holy Spirit, looked intently at him and said, 'O full of all deceit and all fraud, you son of the devil, -And now, indeed, the hand of the Lord is upon you, and you shall be blind, not seeing the sun for a time'. And immediately a dark mist fell on him, and he went around seeking someone to lead him by the hand.*

Acts 19:11-12- "Now God worked <u>unusual miracles</u> by the hands of Paul, so that even handkerchiefs or aprons were brought from his body to the sick, and the diseases left them and the evil spirits went out of them.

Acts 28:3-5- "But when Paul had gathered a bundle of sticks and laid them on the fire, <u>a viper came out because of the heat, and fastened on his hand.</u> So when the natives saw the creature hanging from his hand, they said to one another, 'No doubt this man is a murderer, whom, though he has escaped the sea, yet justice does not allow to live.' But <u>he shook off the creature into the fire and suffered no harm</u>".

Acts 28:7-8-"In that region there was an estate of the leading citizen of the island, whose name was Publius, who received us and entertained us courteously for three days. And it happened that the father of Publius lay <u>sick of a fever and dysentery</u>. Paul went in to him and prayed, and he laid his hands on him and healed him.

Acts 28:9- [Paul had just healed a man with a fever]"So when this was done, <u>the rest of those on the island who had diseases also came and were healed</u>".

So we found that the Heart-brain can send out E-M messages to help heal and the DNA in the hands can send bio-photonic messages to nearby cells to begin healing processes. It happens continuously not because of magic, but because of physics and it has been understood by scientists for some time now as we find more and more out about the truth in the Bible. If the people get really sick and fall asleep, they can not only have the sickness healed, but if one can get the soul to reenter the body, one can revitalize a dead person. It is always better to do this one newly dead people so less of the body must be "repaired".

Revitalization

Beyond healing the sick we need to look at Revitalization. There are several ways that people have come back to life. These include Revitalization, Reincarnation, Possession, and Resurrection. The most direct method which is simply an extension of healing the sick is called Revitalization. The Bible is filled with details about people raising the dead. Here are a few. The information may allow you to understand more about interfacing with dead loved ones, or how free souls may interacted before going to sleep or other such things that have been witnessed.   What we find out is that most of those brought back have no recollection of their death time and they are brought back in the same bodies with the same memories.

1 Kings 17:17-24-- And it came to pass after these things, that the son of the woman, the mistress of the house, fell sick; and his sickness was so sore, that there was no breath left in him. --- [Elijah] he took him out of her bosom, and carried him up into a loft, where he abode, --And he stretched himself upon the child three times, and cried unto the LORD, --And the LORD heard the voice of Elijah; and the soul of the child came into him again, and he revived. [Elijah raised the dead]

2 Kings 4:25-35-- The Shunammite woman --conceived and gave birth to a son.-- something disastrous happened to her son and he died. The Shunammite woman hastily went to the prophet Elisha, who stretched himself upon him: and the child sneezed seven times, and the child opened his eyes.

Kings 13:21-- *And it came to pass, as they were burying a man, that, behold, they spied a band of men; and they cast the man into the sepulcher of Elisha: and when the man was let down, and* <u>*touched the bones of Elisha*</u>*, he revived, and stood up on his feet.*

Luke 7:13-15-- *the Lord --touched the dead boy: and they that bare him stood still. And he said, Young man, I say unto thee, Arise. And he that was dead sat up, and began to speak. And he delivered him to his mother.*

Matthew 9:25/Mark 5:42/ Luke 8:55- *- But when the people were put forth, he* [Jesus] *went in, and took her* [a dead girl] *by the hand, and the maid arose.*

John 11:43-44--*And when he thus had spoken, he cried with a loud voice, Lazarus, come forth. And he that was dead came forth, bound hand and foot with graveclothes: and his face was bound about with a napkin. Jesus saith unto them, loose him, and let him go.*

Matthew 27:52-53-- *And the graves were opened; and many bodies of the saints which slept arose, And came out of the graves after his resurrection, and went into the holy city, and appeared unto many.*

Acts 9:36-42-- *Now there was at Joppa a certain disciple named Tabitha, - it came to pass in those days, that she was sick, and died-- Peter put them all forth, and kneeled down, and prayed; and turning him to the body said, Tabitha, arise. And she opened her eyes: and when she saw Peter, she sat up.*

Acts 20:9-12- *a certain young man named Eutychus, being fallen into a deep sleep --fell down from the third loft, and was taken up dead. And Paul went down, and fell on him, and embracing him said, Trouble not yourselves; for his life is in him. And they brought the young man alive, and were not a little comforted.*

2 Corinthians 12:1-4- I [Paul talking about himself dying the 1st time] know a person in Christ who fourteen years ago was <u>*caught up to the third heaven*</u> *[Died]. Whether it was in the body or out of the body I do not know - God knows. this person was caught up to paradise.*

Acts 14:19-20- Jews from Antioch and Iconium, -- stoned Paul, and drew him out of the city, supposing he had been dead. As the disciples stood round about him, he rose up, and came into the city: and the next day he departed with Barnabas to Derbe [50 miles away]. [Paul was probably dead a 2nd time as people simply left his carcass, then he hopped up and walked 50 miles. He died a third time in Rome and didn't come back.]

I know it sounds like everyone was brought back to life in the olden days, but these instances were rare and controlled. As the people were all "asleep" until Jesus or one of the others pulled them back to life, no one remembered anything about being dead.

Half Way Revitalized

The Bible tells us of another type of revitalization that is temporary and it does not require a dead body. For this we read about the Witch of Endor bringing back the Prophet Samuel for a short visit to talk to King Saul. Please notice bringing people back to life did not require Faith in Jesus.

1 Samuel 28:11-15-Then the woman asked, "Whom shall I bring up for you?"-"Bring up Samuel," he said. When the woman [witch of Endor] saw Samuel, she cried out at the top of her voice and said to Saul, "Why have you deceived me? You are Saul!" The king said to her, "Don't be afraid. What do you see?" The woman said, "I see a ghostly figure coming up out of the earth." "What does he look like?" he asked. "An old man wearing a robe is coming up," she said. Then Saul knew it was Samuel, and he bowed down and prostrated

192

himself with his face to the ground. Samuel said to Saul, "Why have you disturbed me by bringing me up?" Samuel had been "sleeping" and the Witch of Endor had awakened him. He was plenty mad about the whole thing. Certainly, he was not in Heaven and he had not been conscious until Saul had him summonsed. What happened here? --- Samuel's Soul vibration was slowed so that he could experience this world for a time. Yes; he was mad, but that was because he didn't like Saul anymore. This was not sustained and he went back to "Sleep". While this sleep might be until the final resurrection, some have been reincarnated.

From a Fish

Jonah 2:1-6-[Jonah said after coming out of a whale that was underwater for 3 days] From the depths of the grave I called for help, and you listened to my cry. --I have been banished from your sight; yet I will look again toward your holy temple.--- To the roots of the mountains I sank down; the Earth beneath barred me in forever. But you brought my life up from Sheol. [This one is a little different than the others in that God revitalized him, but I think it still belongs here.]

Multiple Revitalizations

As everyone sleeps until the time of God's second coming, possibly hundreds of people were revitalized as their soul that was sitting around came in contact with a spirit and became alive as their body somehow became whole as well when Jesus raised himself from the dead.

Matthew 27:51-53 At that moment the curtain of the temple was torn in two from top to bottom. The earth shook, the rocks split and the tombs broke open. The bodies of many holy people who had died were raised to life. They came out

193

of the tombs after Jesus' resurrection and went into the holy city and appeared to many people.

Can you imagine these people coming back to life. Many texts tell us they had no recollection about what had happened to them as they would have felt they had just now died and came alive immediately. The Bible doesn't say much about them besides this, but we can imagine that they had decomposed some and Jesus would have had to do significant repair on the bodies. Except for the last two examples, normal people with a little more "faith" than most of us used normal physical characteristics to extend someone's life. Besides this capability, the Bible tells us about something called prophecy which initially seems more like magic.

Prophecy

Another simple test is to test prophetic capabilities. While many times prophecy was described as coming from God or provided in a dream or provided interpretation of a dream, the following verse shows that this feat was done by faith, but not the special faith in God all the time. Sometimes it was even controlled in a person by something called demonic possession that we will get to in a bit. The main this needed here is for the future to have already happened as we now know both forward and backward time MUST both exist or time would eventually be gone as it slips away from our universe. To sense the future one must only sense the backward time of our adjoined universe as the future would now be the past. I know it's confusing, but the Bible is not the only characterization of this "normal" capability.

Acts 16:16-18 -"Now it happened, as we went to prayer, that a certain slave girl possessed with a spirit of divination met us, who brought her masters much profit by fortune-telling. This girl followed Paul and us, and cried out, saying, 'These men are the servants of the Most High God, who proclaim to us the way of salvation'. And this she did for many days. But Paul, greatly annoyed, turned and said to the spirit, 'I command you in the name of Jesus Christ to come out of her'. And he came out that very hour". Here is the odd part in this capability. Somehow the possessed person was programmed to sense prophecies or the demon was doing the prophecies. Both possibilities are reasonable with respect to our new found science.

Acts 11:28- *"Then one of them, named Agabus, stood up and showed by the Spirit that there <u>was going to be a great famine throughout all the world</u>, which also happened in the days of Claudius Caesar".*

Acts 21:10-13- *"And as we stayed many days, a certain prophet named Agabus came down from Judea. When he had come to us, he took Paul's belt, bound his own hands and feet, and said, thus says the Holy Spirit, 'So shall the Jews at Jerusalem bind the man who owns this belt, and deliver him into the hands of the Gentiles'.*

There are many, many more, but let's go to some outside of the Bible that show this capability is not a religious capability. Certainly, we must go away from the debased self, sex, survival concentration to allow for this viewing, and many prophecies of the Bible were "given by God" but many were not and they were just as accurate as we also find below.

Proof Outside the Biblical Teachings

In the 15th century 2 great prophets arose to tell us about the future. One was a man named Nostradamus and the second was a woman simply named Mother Shipton. All their predictions come true over time. Mother Shipton who was born Ursula Southeil (1488–1561), told about submarines, women wearing pants and having short hair, Automobiles, the rise of the Church of England, Radios, telephones, telegraphs, hydroelectric power, manufacture of mountain tunnels, airplanes and commercial air travel, & iron ships, the California gold rush, World War I, US Civil War, the French Revolution, Airborne military and their use, British and French alliance during the World War, the Allies and Communist bloc, and the cold war, the France to England underwater tunnel, assemblies would be put together with huge machines, the printing press and how it would change

196

writing forever, and she has given us a very good image of what is to come in our very near future.

Most people know about Nostradamus, born as Michel de Nostredame (1503- 1566), but he predicted Hitler's reign and called him by the name of Hister; he called out events associated with and correctly named Franco; he called out Mussolini's reign and alliance with Germany. He indicated that twin towers would be attacked from the sky that occurred September 11, 2001 in New York; he named and dated the great fire of London; he had many predictions concerning Napoleon which came true; and many, many, many more including, like Mother Shipton, a very detailed account of our very near future. These are true prophetic images not accomplished by heaven sent imagery, per se. They certainly were accomplished by faith, but it is a different type of faith.

Neither of these individuals were deeply religious, but they still were able to see the future in extremely vivid detail outside of religion. That brings us to a disturbing characteristic for some that deals with our bodies stopping its operations. We call it death, but remember only the "self" dies as the spirit portion goes back to the linked universe and the Soul is held in this universe to counter the spirit. Generally speaking while the soul is in "Limbo" it [I mean he or she] is sleeping.

Dead People Sleep

This whole concept of people's souls sleeping when we die should also be investigated. Let me first go through the normal characteristic and then I will talk about a somewhat strange phenomenon that involves a soul waking up after death..

John 11:11-44-He said, and after that He said to them, "Our friend Lazarus sleeps, but I go that I may wake him up." Then His disciples said, "Lord, if he sleeps he will get well." However, Jesus spoke of his death, but they thought that He was speaking about taking rest in sleep. Then Jesus said to them plainly, "Lazarus is dead. --- Jesus, again groaning in Himself, came to the tomb. --- He cried with a loud voice, "Lazarus, come forth!" And he who had died came out bound hand and foot with grave clothes, and his face was wrapped with a cloth. Jesus said to them, "Loose him, and let him go."
Here is something important about this verse besides being revitalized. Lazarus could not tell people about the afterlife, because he had not lived one. He had been "as if asleep". Jesus told everyone that death was like sleep. I don't mean a dreaming sleep. I mean a nothingness sleep.

1 Corinthians 15:50-53- Now this I say, brethren, that flesh and blood cannot inherit the kingdom of God; -- Behold, I tell you a mystery: We shall not all sleep, but we shall all be changed- at the last trumpet - the dead will be raised incorruptible, and we shall be changed. The interesting part of this verse is not that we sleep when we die, but that sometimes, some don't we will get into this later with something called reincarnation.

1 Corinthians 15:40-44-- Behold, I tell you a secret: we shall not all sleep, but we shall ALL BE CHANGED, IN, MOMENT, in the twinkling of an eye, AT THE LAST TRUMP: for the trumpet shall sound, and the dead shall be raised incorruptible, and we shall be changed. Again, no question about everyone "sleeping" when they are dead until Jesus comes back. Everyone who is dead "in Christ" get yanked out of their graves and become immortal. The idea that everyone stays asleep in their graves until Jesus comes back is reinforced in "John", "Ecclesiastes" and "Daniel.

*Ecclesiastes 9:5-*The soul, the living know that they shall die, but the dead know not anything. They are asleep.

*Psalm 90:3-6 -*You turn man back into dust And say, "Return, O children of men." For a thousand years in Your sight Are like yesterday when it passes by, Or as a watch in the night. You have swept them away like a flood, they fall asleep; In the morning they are like grass which sprouts anew.

Job 14:10-12 -"But man dies and lies prostrate Man expires, and where is he? "As water evaporates from the sea, And a river becomes parched and dried up, So man lies down and does not rise. Until the heavens are no longer, He will not awake nor be aroused out of his sleep.

Daniel 12:2 -"Many of those who sleep in the dust of the ground will awake, these to everlasting life, but the others to disgrace and everlasting contempt. The dead sleep until the 2nd coming of Jesus.

*Revelation 14:13 -*And I heard a voice from heaven, saying, "Write, 'Blessed are the dead who die in the Lord from now on!'" "Yes," says the Spirit, "so that they may rest from their labors, for their deeds follow with them."

John 5:28-29 -"Do not marvel at this; for an hour is coming, in which all who are in the tombs will hear His voice, and will come forth; those who did the good deeds to a resurrection of life, those who committed the evil deeds to a resurrection of judgment. Sleep of death lasts until the resurrection at the second coming of Jesus.

John 5:28-29- Do not be amazed at this, for a time is coming when all who are in their graves [sleeping for thousands of years] will hear his voice and come out—those who have done what is good will rise to live, and those who have done what is evil will rise to be condemned.

John 14:2-3- I go to prepare a place for you. And if I go and prepare a place for you, I will come again [in a couple thousand years] *and receive you to Myself; that where I am, there you may be also* [as you will be sleeping until that time.].

Acts 2:29-34-Men and brother, let me speak freely to you, the patriarch David, he is both dead and buried and his tomb is with us to this day.---For David did not ascend into the Heavens. Man oh man! "Why be Christian?" some would say, but Paul states it in the best way. "When Jesus comes back all the dead people rise and BEGIN their eternal COMFORT".

1 Thessalonians 4:13-18 -But we do not want you to be uninformed, brethren, about those who are asleep, so that you will not grieve as do the rest who have no hope. For if we believe that Jesus died and rose again, even so God will bring with Him those who have fallen asleep in Jesus. For this we say to you by the word of the Lord, that we who are alive and remain until the coming of the Lord, will not precede those who have fallen asleep.

Long Sleep

200

The next section I call sleep a long time as it tells us that not only are you going to "sleep" for a long time, but also your ancestors are STILL sleeping when you finally die.

2 Samuel 7:12 -*"When your days are complete and* <u>*you lie*</u> <u>*down with your fathers*</u>*, I will raise up your descendant after you, who will come forth from you, and I will establish his kingdom.*

1 Kings 11:43 -*And* <u>*Solomon slept with his fathers*</u> *and was buried in the city of his father David, and his son Rehoboam reigned in his place.*

1 Kings 14:20 -*The time that Jeroboam reigned was twenty-two years; and* <u>*he slept with his fathers*</u>*, and Nadab his son reigned in his place.*

1 Kings 16:6 -*And Baasha* <u>*slept with his fathers*</u> *and was buried in Tirzah, and Elah his son became king in his place.*

1 Kings 22:50 -*And* <u>*Jehoshaphat slept with his fathers*</u> *and was buried with his fathers in the city of his father David, and Jehoram his son became king in his place.*

2 Kings 14:16 -*So* <u>*Jehoash slept with his fathers*</u> *and was buried in Samaria with the kings of Israel; and Jeroboam his son became king in his place.*

2 Kings 15:7 -*And Azariah slept with his fathers, and they buried him with his fathers in the city of David, and Jotham his son became king in his place.*

2 Kings 16:20 -*So* <u>*Ahaz slept with his fathers*</u>*, and was buried with his fathers in the city of David; and his son Hezekiah reigned in his place.*

2 Kings 20:21 -*So* <u>*Hezekiah slept with his fathers*</u>*, and Manasseh his son became king in his place.*

2 Kings 21:18 -And <u>Manasseh slept with his fathers</u> and was buried in the garden of his own house, in the garden of Uzza, and Amon his son became king in his place.

While people can be awakened from death-sleep, generally speaking, dead people sleep until Jesus comes back. To someone who has passed away, there is no sense of time as we understand it as indicated in Luke. While a number of people have clearly been brought back Samuel, Moses, Isaiah, Elijah, Jonah, Job and others there is a strong belief that very few dead people are "disturbed". Some believe there is a larger number of people who "return", but that is a different story. We call this reincarnation and while it is a very strange thing, it in no way violates the details of reasonable scientific research or theory.

Reincarnation

There seems to be too many people if all previous people's souls are still asleep also the whole concept of God only doing good and at the same time being responsible for massive hurricanes killing many children and innocent people. Logic would indicate that souls are reborn in new bodies many times. This is called reincarnation. The whole concept is odd to say the least. It is when a soul reenters a new body. Most generally, it is the body of a baby and this person is almost completely different than the original. His memories are [usually] different, his demeanor might be different, but there are "things" connected with the soul that make him the same. Rather than explaining, it might be better to simply list a few verses that can help here. Afterwards, I will try to put perspective on them. Throughout the New Testament, you are teased about John the Baptist being the reincarnated Elijah. The interesting part is the John [or Elijah] had no memory of being Elijah, but he still had the same feeling, faith, force, and function [maybe some other "f" Alliteration as well] to continue his work. Whenever John/Elijah saw Jesus, grown up, he immediately worshiped him. You would think that John growing up with Jesus would naturally think Jesus was just a guy. Most of the people around Nazareth thought Jesus was a "false prophet" but John/Elijah had no issue because that part of our memory [love, etc.] is not lost. Those loved ones, still feel the same love and there can be a closeness from that memory. We don't understand too much about the soul, and this reincarnation thing doesn't make it any easier to understand, but that does not mean it violates any reasonable scientific determination.

Matthew Chapter 11 and 18-*"For all of the prophets and the law have prophesized until John. And if you are willing to receive it, He [Jesus] is Elijah who was to come." -- 'Why then do the scribes say that Elijah must come first?' But he answered them and said, 'Elijah indeed is to come and will restore all things. But I say to you that Elijah has come already, and they did not know him, but did to him whatever they wished. So also shall the Son of Man suffer at their hand.' Then the disciples understood that he had spoken of John the Baptist."* This shows a strong belief in Reincarnation, so what happened to John the Baptist/Elijah?

Matthew 17:1-13"*After six days Jesus took with him Peter, James and John the brother of James, and led them up a high mountain by themselves. There he was transfigured before them. His face shone like the sun, and his clothes became as white as the light. Just then there appeared before them Moses and Elijah*- So, john, now Elijah came back to life, talking and completely visible, but then he vanish. These guys had been dead for hundreds of years. They had not been to heaven because Jesus told everyone that he was going to make heaven livable so you ask, "Where were they?" Well the next section tells us a little bit concerning Elijah and where he has been.

Isaiah 53-*After the suffering of his soul, he will see the light [or take in the Holy Spirit thing] and be satisfied; by his knowledge my righteous servant will justify many, and he will bear their iniquities.* This is identifying everyone as being iniquitous, promising substantial suffering, presenting justification and understanding by God, and finally this resurrection thing. Please notice that it is talking about the SOUL suffering rather than people suffering. Why do you suppose that is? I'll tell you. The Soul extends beyond one's life.

Isaiah 66-*"From one New Moon [birth] to another and from one Sabbath to another, all mankind will come and bow down*

204

before me, says the LORD. And they will go out and look upon the dead bodies of those who rebelled against me; their worm will not die, nor will their fire be quenched, and they will be loathsome to all mankind." We don't know what else Isaiah might have written for his book ends with this final warning. The rebels mentioned here seem include all those who would not bring in the Holy Spirit. This included those who died well before the time of Jesus. The only way all mankind worship God and view the dead rebels is for most people to take this Holy Spirit in during one of several successive lifetimes allowing for the acceptance.

Learning from the Reincarnated

Daniel 11 and 12-*"Those who are wise will instruct many, though for a time they will fall by the sword or be burned or captured or plundered."* People will be teaching others even after they have fallen by the sword sounds like one can instruct after death. While there might be some way for a dead person to instruct, we can more easily understand how a person can be resurrected and pass on things he learned from previous lives. These would not be memories; they would be important things like understanding, humility, hope, love, charity, God, and empathy. He would instruct those things. *"Some of the wise will stumble, so that they may be refined, purified and made spotless until the time of the end, for it will still come at the appointed time."*

EVEN the wise will "stumble" in one life, but they will have more chances until the end of time. I think this verse is pretty clear. Reincarnation is to allow the most people to be saved for the final resurrection. We will look at the science at the end.

"Multitudes who sleep in the dust of the earth will awake: some to everlasting life, others to shame and everlasting contempt."

205

The verses say, after a number of reincarnations, there will be a resurrection. If, after many stumbles, one finally takes the Holy Spirit, this resurrection will be neat and while you are "living", your vibration level will be faster so you will be able to affect other people in positive ways during your life and AFTER.

Fight Against Entropy

Man cannot fight carnality without the aid of the Holy Spirit and when someone dies without this element, he cannot communicate with the living. Certainly he can be seen or felt just as demons but meaningful interface must have the spirit to bring out a level of life. When the witch of Endor revived the soul of Samuel, it was because of this "spirit". Those without the spirit, evidently have more than one chance. It is not well understood how all this happens, but there are reincarnations and there has been a long delay before the final resurrection for one thing as indicated in the following verses and others.

Matthew 11:14 and 17:12-13, concerning the identity of John the Baptist—Both understood he carried the soul of Elijah.

John 9:2, "Who sinned, this man or his parents, that he was born blind?" The disciples understood that reincarnations carried punishments on occasion.

Revelation 13:10, "If anyone is to go into captivity, into captivity he will go. If anyone is to be killed with the sword, with the sword he will be killed." Again we find that some punishments await the reincarnated.

Hebrews 9:,27-28 Just as man is destined to die once, and after that to face judgment, so Christ was sacrificed once to take away the sins of many people; and he will appear a second time, not to bear sin, but to bring salvation to those who are waiting for him. This is talking about death of the soul which happens unless someone is forgiven as indicated in the Christian doctrine. It is not saying the other reincarnation passages are lies.

God's Return Is At Hand

If this reincarnation/reentry capability to increase our possibility of gaining the holy spirit was not so, the thousands of years God stays away does not make much sense. Jesus continued to tell his disciples that *the time of God's return was at hand*. They thought it was to be in their lifetime. In one way, it was as more and more believers are collected for his final return and we are not aware of any time spent between lives or in previous lives, so to us the time is still at hand.

*2 **Peter** 3:8-9…But do not let this one fact escape your notice, beloved, that with the Lord one day is like a thousand years, and a thousand years like one day. The Lord is not slow about His promise, as some count slowness, but is patient toward you, not wishing for any to perish but for all to come to repentance.*

God wants as many to accept the Holy Spirit as possible.

*1 **Timothy** 2 :3-6 …For this is good and acceptable in the sight of God our Savior; Who will have all men to be saved, and to come unto the knowledge of the truth. For there is one God, and one mediator between God and men, the man Christ Jesus; Who gave himself a ransom for all, to be testified in due time. Certainly there are fewer believers today than during the height of the Christian revolution but Jesus has not returned.* By this the reason is that he wants to give people as many "redos" as possible just in case he can get one more follower.

God does Only Good

We are told in the book of Romans, rightly, that God only does good, but how is killing innocent people with a hurricane or tidal wave considered good if God wants as many to come to him as possible? How could God tell the Jews to kill every man, woman, and baby Amalekite? The answer is that reincarnation can only happen if one dies. To get away from some limitation that would not allow someone to believe and not be part of the second death, many times might require their untimely death a person to gain a redo.

Romans 8:28 -*And we know that in all things God works for the good of those who love him, who have been called according to his purpose.*

The Generation Will Not See Death

Oh Boy! We read the next group of verses and claim this cannot be right. Jesus told his disciples exactly how he would finally return for his second coming and then he told them something strange. He said some would not taste death before he returns and the generation will not pass away. While there is no question that all of the people who were with him had a physical dying well before Jesus' return, some would have been reincarnated according to these passages

Matthew 24:29-34- 'the sun will be darkened, and the moon will not give its light; the stars will fall from the sky, and the heavenly bodies will be shaken.' "Then will appear the sign of the Son of Man in heaven. And then all the peoples of the earth will mourn when they see the Son of Man coming on the clouds of heaven, with power and great glory And he will send his angels with a loud trumpet call, and they will gather his elect from the four winds, from one end of the heavens to the other. "Now learn this lesson from the fig tree: As soon as its twigs get tender and its leaves come out, you know that summer is near. Even so, when you see all these things, you know that it is near, right at the door. Truly I tell you, <u>this generation will certainly not pass away</u> until all these things have happened.

208

Mark 9:1-13-And he said to them, "Truly I tell you, <u>some who</u> <u>are standing here will not taste death</u> before they see that the kingdom of God has come with power." --And they asked him, "Why do the teachers of the law say that Elijah must come first?" Jesus replied, "To be sure, Elijah does come first, and restores all things. Why then is it written that the Son of Man must suffer much and be rejected? But I tell you, Elijah has come, and they have done to him everything they wished, just as it is written about him."

Mark 13:24-28-"But in those days, following that distress, 'the sun will be darkened, and the moon will not give its light; the stars will fall from the sky ,and the heavenly bodies will be shaken.' "At that time people will see the Son of Man coming in clouds with great power and glory. And he will send his angels and gather his elect from the four winds, from the ends of the earth to the ends of the heavens. "Now learn this lesson from the fig tree: As soon as its twigs get tender and its leaves come out, you know that summer is near. Even so, when you see these things happening, you know that it is near, right at the door. Truly I tell you, <u>this</u> <u>generation will certainly not pass away until all these things</u> <u>have happened.</u>

Matthew 16:27-28-For the Son of Man will come in His Father's glory with His angels, and then He will repay each one according to what he has done. Truly I tell you, <u>some who are standing here will not taste death</u> <u>until they see the Son of Man coming in His kingdom.</u>

Scientific and Modern Reincarnation

Science cannot understand death. How something dies is a mystery. I explained life is outside the body and it supply sensory inputs to activate the storage device called DNA. Knowing how to build an entire person, DNA relies on external stimulate to know what body part to place where.

209

The soul supplies the DNA instructions as it is the contact with the world outside the body and outside what we sense as reality.

Our focus on Sex, Self, and survival, push the soul into a level of limited use, but it is still there. When body death occurs, this dimensional element of our universe is released, but it cannot change its energy level without an equal change from our adjoined universe. Therefore it hangs around. While most texts tell us the souls are almost always sleeping. At the moment of inception there is some type of cry that can wake up souls so they can enter the new entity and produce life inside. Here is what we know. There are a finite number of souls and many have died which would rob the world of that many people that could live unless souls can be reused in reincarnation just as described in Judeo-Christian texts.

Parapsychology Science and Reincarnation

Dr. Ian Stevenson, former Professor of Psychiatry at the University of Virginia School of Medicine and former chair of the Department of Psychiatry and Neurology, dedicated the majority of his career to finding evidence of reincarnation, until his death in 2007. During that time he claimed to have found over 3,000 examples of reincarnation which he shared with the scientific community.

Xenoglossy and Reincarnation

Ian Stevenson investigated numerous cases of something called xenoglossy, which is defined as *"speaking a real language entirely unknown to (the speaker) in his ordinary state."* The definition was coined originally by Charles Richet between 1905 and 1907. Richet was a Nobel Prize–winning doctor, whose did about the same thing as Stevenson.

English woman become Swedish Man-One xenoglossy investigation involved a 37-year-old American woman whose initials were TE was born and raised in Philadelphia. Her parents spoke English, Polish, Yiddish, and Russian at home while she was growing up and she studied French while in school. Here is the weird part. While under eight different regression hypnosis sessions, TE became "Jensen Jacoby," a male Swedish peasant. TE answered questions posed in the Swedish language with Swedish responses, using about 60 words not first spoken by a Swedish-speaking interviewer. TE as Jensen was also able to answer English questions with English answers. Polygraph tests and similar tests showed shew was Swedish and the accent was praised by Stevenson's consultants however the language was mixed with Norwegian.

I'm not going into all of these things, but imagine 3000 similar reincarnation studies by this scientist alone. You may have been a female and male a small child dying young or all sorts of things and a little bit of your past is still part of you until our final resurrection. Before that happens, I need to clarify another thing that seems strange in the Judeo-Christian texts that one might call time travel.

Time Travel

If we look at the New Testament closely it seems that no time passes for sleeping souls. We had better see what the Bible tells us and look at the science. Hopefully you still remember that almost all dead people are asleep in the ground and some stay there for thousands of years before God comes back with a new heaven for us to live in. At that time all the dead will rise and some will go up and some will not. Even the reincarnation stuff isn't so weird, but then we come to the thief on the cross who wanted to go to heaven and Jesus said. Here is what Jesus [telling the truth] told him.

Luke 23:43 Jesus answered him, "Truly I tell you, today you will be with me in paradise."

Yes! Jesus did not even rise again to go to his "father" for 3 days and no one is going to heaven for a while so how could he have been telling the man the truth and be in accordance with modern science? If you remember I told you that if you were capable of going the speed of light you would never age. Also you must understand that for you no time would pass during you speed of light journey. If you were on this journey for 3 thousands years, your hair would not grow, you would have no time according to modern relativistic science. Then I told you that you could not go that fast because what made you a "you" would be changed by the event. The soul is different. While no time would pass, if dead souls simply vibrated at the speed of light, a thousand years would be like a day and the thief could go to paradise with Jesus the same day. Here are some more time travel verses.

2 Corinthians 5:8-We are confident, I say, and willing rather to be absent from the body, and to be present with the Lord.

2 Peter 3:8-9...But do not let this one fact escape your notice, beloved, that with the Lord one day is like a thousand years, and a thousand years like one day.

These verses were not to add confusion but to let you know when your science was getting it right. Unfortunately there is one more thing that we must discuss and that is demonic possession. Some would simply indicate it is fanciful and others might say it happened during the time of the Romans but not today and others would try to separate what they believed to be science from religion to allow for this awful but important group of passages.

Demonic Possession

The New Testament is filled with examples of demons and demons possessing people. We had better see what is said and how it plays with science. We live in a world that allows so much tolerance for moral deviants, for insanities of action, for all the things demons can make people do that we don't even see the thousands or millions of demonic possessions that the Bible warns us about. It also tells us the simple name of Jesus can drive out unwanted demonic souls. While demons are not the cause of all insanities, hate, killing, moral debasement and all of the rest, if we ignore them they will continue to gain power over people. For this section I think I had better go back to the beginning and fill in a few holes so it makes sense. During an ancient time there was a horrible war in Heaven between angels who wanted to live more carnal, debased, deviant lives and those would follow God's commands and act uprightly and feed on love rather than hate. About 1/3 of the angels wanted this "more free society". As the war ended, the rebels who wanted carnal life lost. OK! The creator of the universe was on the other side. As punishment, the rebel Angels became human again, but with no spirit [or light]. Without this spirit, they would never be able to leave the carnal universe and they would have to roam the land as a demon after death until the second coming of Jesus. Here is one of many texts concerning the loss of their spirit.

214

Jubilees 2:9-Nor may we take revenge on him because he has stripped us of the "light". He marked out the borders of the world and created man in his own image with whom he hopes again to people heaven, with pure souls.

These people were called the Nephilim or Anak by the Jews, Annunaki, by the Sumerians, Archaics [Adena-Ohio Valley], Akamim [Mongulala-Brazil], Lords of Amenti [Khemetian-Egypt], Arya [Dravidian-India], and the Olympians by the Greek. Many were worshipped as gods as they had the ancient knowledge and they were giant people. After a time, they began to die off, in fact by about 1000BC all of these people and their descendants were wiped off the Earth, but they could never leave so they had to exist in a state of limbo and are called demons. The following diagram may help describe the beginnings and endings of these people who became demons.

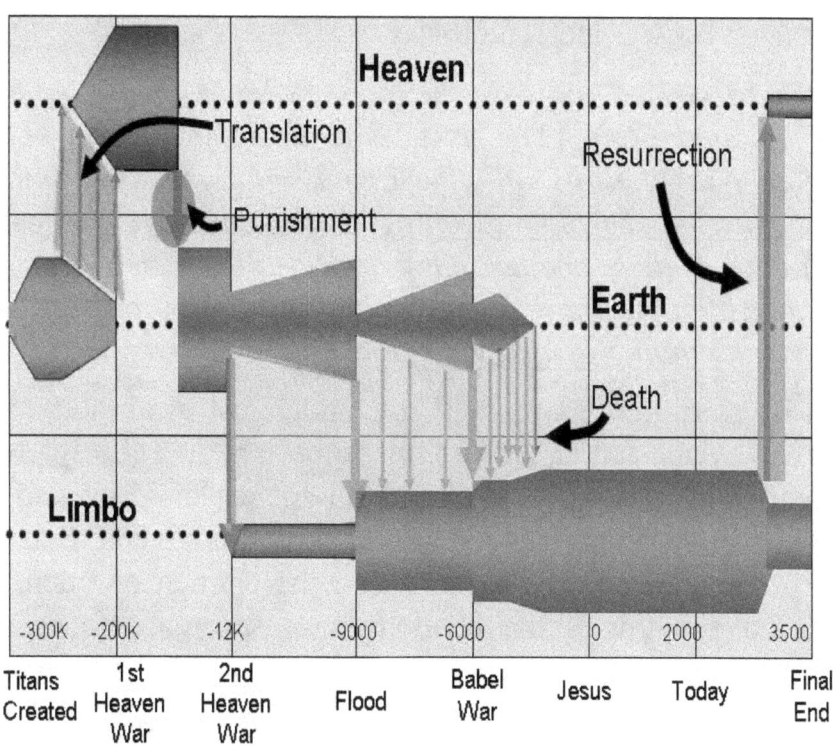

215

Notice all angels [except Archangels] were once normal people who died and went to heaven [the top dotted line]. Forced back to earth, they are shown along the middle line. We can believe there were hundreds of thousands are even millions of these unfortunate people. The lower dotted line is this limbo place. Let me tell you how bad limbo is. While a number of demons had taken possession of a live person so they could get away from the horrors of limbo [no reality— just existence]; when Jesus drove them out the begged to be allowed to enter pigs rather than go back to their punishment place that lasted for centuries. Please understand this if you don't understand anything else about this book; demons will do anything to get out of limbo and experience life for a time. I did place a small line from limbo to heaven at the end as there are some texts that indicate even demons may have hope, but I really don't know about how many. Let's read what the canonized book Enoch had to say.

Enoch 15: 8-12- *"And now, the giants, who are produced from the soul and flesh,* [This is talking about the losers of the Heaven War] *shall be called evil souls upon the earth, and on the earth shall be their dwelling. Evil souls have proceeded from their bodies; because they are born from men and from the holy Watchers is their beginning and primal origin; -And these souls shall rise up against the children of men."*

Yes, I was right; demons will rise up against us all and do all the time. Many fail to discuss this important and deadly group of dead souls described in almost all ancient histories and the Bible. Without them we have no demons, we have no Satan to worry about, we have no serpent in the Garden of Eden, we don't have many of the challenges of spiritual life as was expressed by Paul.

Before the Anak people died, they were the uncontested rulers around the world. After they died, "normal humans" took control of their own fate, but demons were still lingering in the

216

shadows. Paul described them by their level of leadership in the "Limbo place". Demonic Thrones, demonic Principalities, demonic Powers, demonic Authorities, and simply Demons are described, but they are all the wandering souls of Anak people. Luckily, we can be protected. If you aren't frightened, please understand that 1/3 of all the inhabitants of Heaven became demons. We can believe that to be millions.

Romans 8:37–39: "- I am convinced that neither death nor life, neither angels nor demons, neither the present nor the future, nor any [demonic] powers, neither height nor depth, nor anything else in all creation, will be able to separate us from the love of God that is in Christ Jesus our Lord."

Invasion and Removal

The Bible and other documents describe powerful demons that invade people to possess them and that one can demand the demons to leave. New sciences help us understand this is possible as the Soul is essentially a separate entity that co-lives with our Self and a second Soul would naturally be able to make his way into that union.

Matthew 9:33-34: And when the demon was driven out, the man who had been mute spoke. The crowd was amazed and said, "Nothing like this has ever been seen in Israel." But the Pharisees said, "It is by the prince of demons that he drives out demons."

Matthew 10:8: Heal the sick, raise the dead, and cleanse those who have leprosy, drive out demons. Freely you have received, freely give.

Matthew 17:18: Jesus rebuked the demon, and it came out of the boy, and he was healed from that moment.

Mark 1:34: and Jesus healed many who had various diseases. He also drove out many demons, but he would not let the demons speak because they knew who he was.

217

Mark 1:39: So he traveled throughout Galilee, preaching in their synagogues and <u>driving out demons</u>.

Mark 3:15: and to have authority to drive out demons.

Mark 6:13: They <u>drove out many demons</u> and anointed many sick people with oil and healed them.

Mark 16:17: And these signs will accompany those who believe: In my name they will <u>drive out demons</u>; they will speak in new tongues;

Luke 4:33: In the synagogue there was a man possessed by a demon, an evil spirit. He cried out at the top of his voice, ". . . Be quiet!" Jesus said sternly. "Come out of him!" Then the demon threw the man down before them <u>all and came out</u> without injuring him.

Luke 4:41: Moreover, <u>demons came out</u> of many people, shouting, "You are the Son of God!" But he rebuked them and would not allow them to speak, because they knew he was the Christ.

Luke 8:2: and also some women who had been cured of evil spirits and diseases: Mary from whom <u>seven demons had come out</u>;

Luke 8:35: and the people went out to see what had happened. When they came to Jesus, they found the man from whom the <u>demons had gone out</u>, sitting at Jesus' feet, dressed and in his right mind; and they were afraid.

Luke 9:1: When Jesus had called the Twelve together; he gave them power and authority <u>to drive out all demons</u> and to cure diseases,

Luke 9:42: Even while the boy was coming, the demon threw him to the ground in a convulsion. But <u>Jesus rebuked the evil spirit</u>, healed the boy and gave him back to his father.

Luke 10:17: *The seventy-two returned with joy and said, "Lord, <u>even the demons submit</u> to us in your name."*

Luke 11:14: *Jesus <u>was driving out a demon</u> that was mute. When the demon left, the man who had been mute spoke, and the crowd was amazed.*

John 10:21: *But others said, "These are not the sayings of a man <u>possessed by a demon</u>. Can a demon open the eyes of the blind?"*

1 Corinthians 10:20: *No, but the sacrifices of pagans are offered to demons, not to God, and I do not want you to be participants with demons. . . <u>You cannot drink the cup of the Lord and the cup of demons too</u>; you cannot have a part in both the Lord's Table and the table of demons.*

1Timothy 4:1: *The Spirit clearly says that in later times some will abandon the faith and <u>follow deceiving spirits and things taught by demons.</u>*

James 2:19: *You believe that there is one God. Good! <u>Even the demons believe that--and shudder.</u>*

Revelation 9:20: *The rest of mankind that were not killed by these plagues still did not repent of the work of their hands; they <u>did not stop worshipping demons</u>, and idols of gold, silver, bronze, stone and wood -- idols that cannot see or hear or walk.*

Revelation 16:14: *<u>They are spirits of demons</u> performing miraculous signs, and they go out to the kings of the whole world, to gather them for the battle on the great day of God Almighty.*

Revelation 18:2: *With a mighty voice he shouted: "Fallen! Fallen is Babylon the Great! She has become a <u>home for demons and a haunt for every evil spirit</u>, a haunt for every unclean and detestable bird.*

How to Remove Demons

We are told that with Faith in God's Power we can push out Demons, but also this can be accomplished with a type faith that is the separation from carnal thought.

Mark 9:38-40 -Now John answered Him, saying, Teacher, we saw someone who does not follow us casting out demons in Your name, and we forbade him because he does not follow us. But Jesus said; Do not forbid him, for no one who works a miracle in My name can soon afterward speak evil of Me. For he who is not against us is on our side.

ACTS 9:13-14 *Some Jews who went around driving out evil spirits tried to invoke the name of the Lord Jesus over those who were demon-possessed. They would say, "In the name of the Jesus whom Paul preaches, I command you to come out." Seven sons of Sceva, a Jewish chief priest, were doing this.*

Using this less than perfect "faith" and simply saying the name of Jesus did not always end well. Let's continue the story about the sons of Sceva.

ACTS 9:15-16 *One day the evil spirit answered them, "Jesus I know, and Paul I know about, but who are you?" Then the man who had the evil spirit jumped on them and overpowered them all. He gave them such a beating that they ran out of the house naked and bleeding.*

Science and Evidence

Demons and demonic possessions have been seen for centuries. There is no religious limitation concerning this. People are infected and many go to their graves infected. Don't get me wrong here. I am not talking about someone who is a Schizophrenic or Sociopathic killer, however, it must be remembered that there may be a million or more of these desperate souls pushing their way into our reality. I would have to say the really big targets for these demons are the world leaders who may start reasonable, but soon they are destroying their own people, practicing genocide and torture. Remember any expanded carnal action would greatly enhance the possession for a demon so they would focus on those who could obtain power the most. I'm not getting into the apparently possessed leaders that have done these things for no reason beyond giving their possessor pleasure.

As far as science indicating that 2 souls could operate a single body. What we do know is that the souls can temporarily leave for a time. Called by different names astral projection has been studied and even trained over the centuries. Possibly we can even see the image of a soul leaving the body now that cameras have extended wavelength processing and the souls would have some type of wavelength or vibrational component. Here are a couple that were both recorded in 2014 from security cameras. The first one is of a man sleeping in a bed. After some time the security camera say something almost unbelievable a ghostly apparition comes out of his body and flies away.

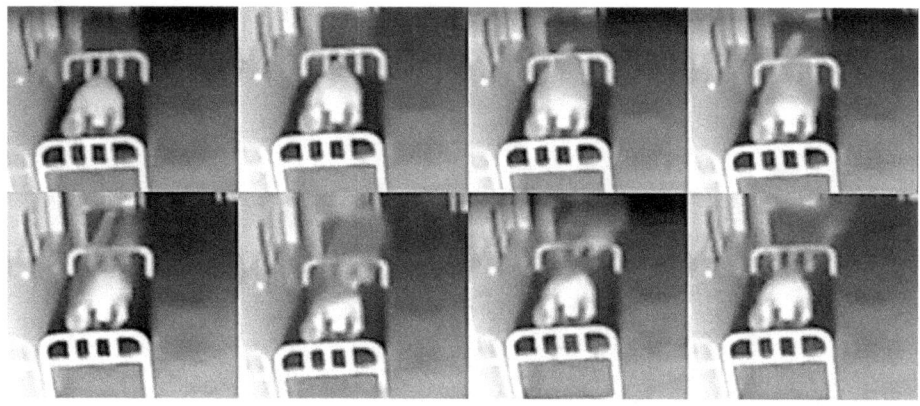

It is not known if this was a demon soul or the soul of the person in the bed, but it does show that souls can leave a live body for a time. The second one shows a shadowy image at one end of a hallway in a hospitol. Within a period of 2 tenths of a second the apparition quickly goes along the hallway to the other side. Both of these come from video streams that may allow an eaisier viewing, but I think these frames show what I'm talking about [a walking or flying unfettered soul]. This could be the soul od an alive person, recently dead person or one of the demonic souls I have been talking about.

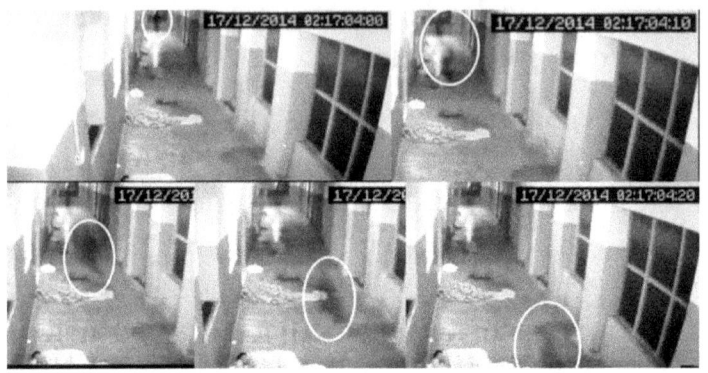

Conclusions

Hopefully you have a better appreciation of how the Bible actually helps prove scientific discoveries and many scientists would be better off to simply start with the Bible rather than trying to disprove it. There is a lot more to say, but I'm afraid that if I say the word Physics once more you will throw the book and someone might get hurt. Hopefully you at least got something out of this book like the things listed.

- **Healing** is not just about getting the right drugs to numb parts of your body or whatever. It has been said that medical treatments are the 1/3 highest cause for loss of life. Both your heart and hands hold a healing power not to be ignored. Luckily today we are being forced to reexamine the methods described by Jesus.

- **Modification of reality by thinking** is not only possible, it has been demonstrated over and over and the participatory physics of today shows Jesus was telling the truth that you can move a mountain with "faith" the size of a mustard seed.

- **Walking on Water-** Jesus was almost mad that his disciples did walk on water, but even Canadians have made things like bowling balls have no weight so that they would not sink. I was as if Jesus knew things we did not.

- **Men during the time of Dinosaurs-**proves that on the 6th day God made man to REPLENISH the earth.

- **The creation of Homo-Erectus-** describes the 6th day man perfectly.

- **The Creation of Cro-Magnon-** greatly compares to the Biblical Adam.
- **The Pleistocene War-** Described in the Judeo-Christian histories, scientists proved it by describing the "Young Dryas", the massive increasing radioactivity of the Earth, and huge amounts of mutation in humans.
- **The Destruction of Venus-** The destruction of Rahab is well documented in Judeo-Christian texts and science, and physical evidence is showing the Bible was right.
- **The Shift in the Earth's Axis-** Hidden from children for some reason, this was not only known by the ancient Jews and the rest of the world, but also modern scientific details keep coming back to it.
- **The Bharata War-** The Bible called it the destruction of the "Tower of Babel" and the seemingly strange outcome of not understanding as much as we once did. Other Judeo-Christian documents filled us in on the horrors as all nations of the world celebrated its ending. It resulted with massive human changes in human DNA, brain size, and even length of life. Scientists are puzzled by anomaly, Historians leave the entire event out, and schools seem to ignore this horror. We should remember and revel in the details of our Bible.
- **Reincarnation-** Today thousands of people remember things they could not remember. Idiot savants play the piano like they knew already; regressionists know a tiny piece of the details that our Bible describes as so important for God to allow conversion of as many people as possible to be with him after death.
- **Demonic possession-** The New Testament pleads over and over to be aware of principalities, powers, and all the other demons that are desperate to regain life in any way they can. We live in a world today that accepts almost any

debauchery or deviant lifestyle we can't even recognize the signs of what was warned over and over.

- **Resurrection**- Science has proven Heaven, the power of the Soul that lives forever, and God who controls existence. If we simply follow God's yearnings resurrection will be to the serenity, love, communion and eternity of Heaven. Science also tells us there are other universes. The Bible tells us of one universe cannot hold a "reality" where you could never communicate with others, where there is no light, where there is only agony. Like a fiery furnace there is only existence. Heaven is better.

The End

About the Author

Steve Preston is a long lime author of scientific, esoteric facts. His books focus on the painful truths rather than whitewashed details that make us comfortable. If you are interested in the truth instead of comfort, please review other works by Mr. Preston as shown below. The images are some from Egypt taking the older version of taxi. To the right the writer is shown in the Jewish Negev desert of Israel where the Dead Sea Scrolls were found. I found nothing but the marvels of Egypt and Israel.

Searching at Egyptian Pyramid   Searching in Israeli Negev

Development of Mankind Series
A Closer Look at Ancient History   *Man After The Flood*
A New View of Modern History   *The Creation of Adam and Eve*
The First Creation of Man   *Six Deaths of Man*
The Second Creation of Man   *Twentieth Century to the End of Time*
The Antediluvian War Years
Bible Series
Adam to Abraham   *Moses to Jesus*
Abraham to Moses   *Understanding the New Testament*
Bible Enhancements
Adam's First Wife   *Expanded Genesis*
Closer Look At Genesis   *Incarnations of God*
Errors in Understanding   *New look at the Bible*

The Devil

The Antichrist

Why the King James Bible Failed

Tracing Cro-Magnon to Jesus

Old Testament Used By Jesus

Sex Crazed Angels

American Current Affairs

Allah' God of the Moon

Promote the General Welfare

Make Your Own Global Warming

Terror of Global Warming

Can We Save America?

Contemporary Issues

American School Disaster

Great American Quiz

Fast History of MILES Training

Our Very Odd Presidents

Monsters are Alive

The Bad Side of Lincoln

Vampires among Us

Consensus Science

America's Civil War Lie

Humans on Display

Ancient History

Anakim Gods

Kingdoms Before the Flood

Ancient History of Flying

Lizard People

Behind the Tower of Babel

World War Before

Driven Underground

World War with Heaven

Four Armageddons

World War Zero

History Confirmed By The Bible

When Giants Ruled the Earth

Anthropic and Esoteric Science

Anthropic Reality

True Happiness

Awaken the Departed

Self-Virtualization

Life Resonance

Self, Soul, Spirit

Releasing Your Consciousness

Biophotonics and Healing

UFO, Astronomy, and Earth Science

Complex Earth

Not from Space

Creation and Death of Dinosaurs

Where UFOs Go

Retiming the Earth

Living on Venus

Victory of the Earth

When Earth Exploded

Martians

Anthropology and Physical Science

DNA of Our Ancestors

Walk Through Time or a Wall

God Didn't Make The Ape

Meaning of Life and Light

Races of Men

Mystery of Photons and Light

Our 12-Dimensional Universe

Slip Through a Wall

Is Time Travel Possible?

Of Science and Religion

Vibrational Matter

Irish-Jewish-Egyptian-Phoenician-Roman Connection

Mysteries of the Exodus

Moses Saved Egypt

Mysterious Pyramids

Why Rome Fought the Berserkers

Scythians Conquer Ireland

Truth About Hyksos Pharaohs

Truth About Phoenicia